Unearthing the Ravages of Informality to Model Resilient Cities Against Climate Change and Ecological Disturbance

Resilient Cities in a World of Informality, Climate Change and Ecological Discrepancy

Unearthing the Ravages of Informality to Model Resilient Cities Against Climate Change and Ecological Disturbance

Resilient Cities in a World of Informality, Climate Change and Ecological Discrepancy

Mululu Chirwa

ELIVA PRESS

ISBN: 978-99949-8-661-3

Cover Design: Eliva Press
Cover Image: Ingimage
Email: info@elivapress.com
Website: www.elivapress.com

This Eliva Press imprint is published by the registered company Eliva Press Global Ltd. part of Eliva Press S.R.L. Publishing Group

The registered company address is: Pope Hennessy Street Level 2, Hennessy Tower Port Louis, Mauritius
Eliva Press S.R.L. Publishing Group address is: Bulevardul Moscova 21, Chisinau, Moldova, Europe

A NEW URBAN DESIGN

UNEARTHING THE RAVAGES OF INFORMALITY TO MODEL RESILIENT CITIES AGAINST CLIMATE CHANGE AND ECOLOGICAL DISTURBANCE

Resilient Cities in a world of Informality, Climate Change and Ecological Discrepancy

By

Mululu Chirwa

(mululuchirwa@gmail.com)

Eliva Press

2023

TABLE OF CONTENTS

LIST OF FIGURES

LIST OF TABLES

LIST OF ACRONYMS

AI: Artificial Intelligence

ANOVA: Analysis of Variance

CBD: Central Business District

CFC: Chlorofluorocarbon

FRA: Forest Resource Assessment

GDP: Gross Domestic Product

GNP: Gross National Product

HDI: Human Development Index

HHRC: High Heat Retention Capacity

HHP: Hedonic House Pricing

JICA: Japan International Cooperation Agency

LCC: Lusaka City Council

NGO: Non-Governmental Organization

NRDC: Natural Resource Development Center

NRC: National Research Council

pH: Potential Hydrogen

ppb: Potential Per Billion

ppm: Potential Per Million

PPPHPZ: Poor People's Process on Housing Poverty in Zambia

PPS: Polycentric Planning System

SDG: Sustainable Development Goal

UN: United Nations

CHAPTER ONE: INFORMAL LAND MARKET DYNAMICS IN KALINGALINGA SETTLEMENT OF LUSAKA

Abstract

Informal land markets characterize most improvement areas in the global south and are evident in Kalingalinga settlement of Lusaka. After physical upgrade Kalingalinga has increased in value over time. Old residents are selling their properties and relocating to places where there is cheaper land. The work explores the reasons for sale of properties, sellers' places of relocation, and quality of sellers' lives after the sale of properties in Kalingalinga. 20 key informants and five Kalingalinga residents were sampled purposively, while 30 sellers who had relocated were sampled through snow ball, adding up to a total sample size of 55 respondents. Findings of the study show that 41.8% sold properties due to family disputes and 30.1% due to poverty. Family dispute emerged as the prominent reason for the sale of properties. Other reasons in their order of decent are: attractive prices; crisis; rising property values; desire to improve life; neighbourhood influence; land use change; ignorance on leasing, rise in cost of living, and conflict in land use. Also finding reveal that, the most important places of relocation are Mtendere East (52.7%) and Kalingalinga (the area further away from Alick Nkhata stretch) (40%). Other places of relocation in order of descent include: Mtendere; NRDC; Kabangwe; farms; Chongwe; outskirts; Ngómbe; Chalala; Kabanana; Mumbwa; and Kafue. Most places of relocation had a background of informality and were further away from the CBD. The study used a paired sample t-Test at alpha level 0.05 (2-tailed) and frequency charts to assess the sellers' quality of life before and after the sale. Findings reveal that most houses acquired after selling were cheaper, had fewer rooms, lacked water and electricity, but had better building materials and toiletry. They were located far from workplaces, town centre, main market, and health centres. Although most Lusaka City Council (LCC) officials said that the sellers benefited, some Poor People's Process on Housing Poverty in Zambia (PPPHPZ) officials, and the sellers themselves said that they lost out. The study concluded that much of the implications of informal land markets on the sellers were negative thereby creating a need for further assessment on how institutions can realign informal land markets to establish a win-win situation between buyers and sellers.

***KEY WORDS**: Informal Land markets, Gentrification, Kalingalinga*

1.1 INTRODUCTION

Informal land market is the selling and purchasing of land/residential properties without the engagement of local authorities (UN-Habitat, 2010 In Anabaraonye, 2017; Thirkell, 1996; Berrisford, 2007). Informal land markets exist in many informal settlements. Thirkell (1999) and Berrisford (2007) were particularly concerned about winners between sellers and buyers. Their case studies in Cebu city and Johannesburg showed that sellers lost out because the transactions are done in their desperate condition. Other scholars add that, informal land markets disadvantage sellers since they later buy properties of lesser value than the ones they sold (Taher and Ibrahim, 2013; Marx, 2010 Anigacz, 2010; Tinsley, 1997). Such a concern for Kalingalinga where informal land markets are taking place motivated this study. The study measures quality of the seller's lives in their places of relocation. Feldman (2014) indicated in his case study of New York that sellers end up in more squalor than before selling their properties. He found that the places of resettlement where congested, and devoid of basic necessities like descent housing, water supply, drainage and sanitation, and electricity.

As Kalingalinga experiences rapid sell of residential properties, buyers redevelop old infrastructures for business while the sellers often relocate to other low-cost residential areas. Little is known about where the sellers go and what happens to their quality of life. This study will provide local authorities and NGOs with information that will be helpful in sustainable settlement upgrading and inclusive transition of land use change in Kalingalinga settlement. The study is a build-up on the dissertation by Chikuta (2016) which claims that the sellers of residential properties in Kalingalinga relocate to more depraved areas and their livelihood degenerate. While he claims that the sellers relocate to places of depravity, his study does not follow up the sellers in their places of relocation to assess their living conditions. This study therefore follows up the sellers in their places of relocation to assess whether their living conditions have improved or declined since they sold their property from Kalingalinga. The study seeks to address the following research questions and hypothesis:

i. Why do Kalingalinga residents sell their residential properties?

ii. Where do sellers relocate after selling their residential properties in Kalingalinga?

iii. Does the sellers' quality of life improve or decline after the sale of residential properties in Kalingalinga?

iv. Null Hypothesis (Ho): there is no significant difference in the quality of life for property seller in Kalingalinga before and after the sale.

Alternative Hypothesis (Ha): there is a significant difference in the quality of life for property sellers in Kalingalinga before and after the sale.

1.2 LITERATURE REVIEW

The study is based on economic land rent theory which more than often rallies alongside urban gentrification. Urban gentrification happens in a locality where the forces of investment overrule a poor community where residents fail to cope with the rise in cost of living due to increased land value, and the temptation to sale houses is greater than that of holding on (Anabaraonye, 2017). The problem is that the sellers often relocate to other informal areas which are places of environmental depravity characterized by sub-standard housing. Figure 1. shows a summary of the theoretical framework for the study.

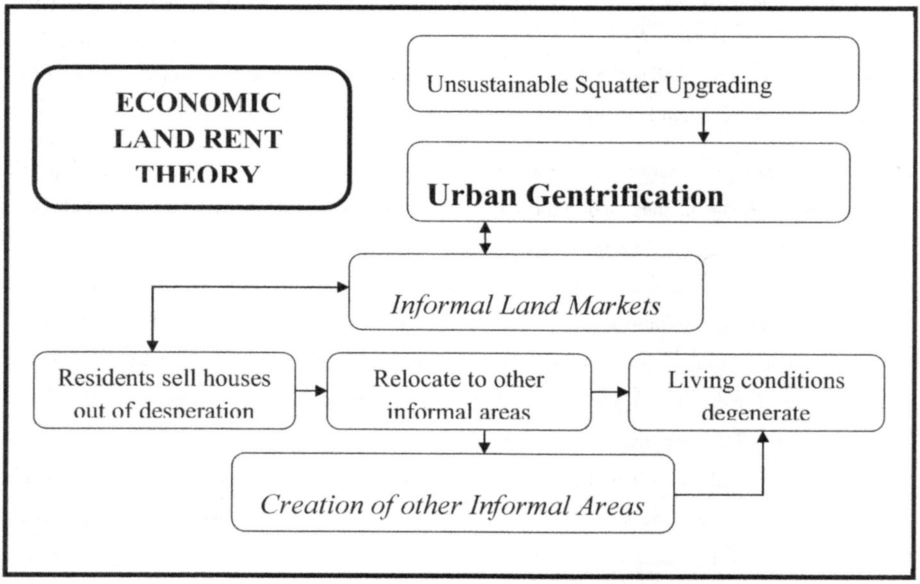

Figure 1. Showing The Theoretical Framework of the study

Source: Generated by Author, 2017

Productive improvements on land increase its value (Ibrahim and Taher, 2014; Mohammed *et al.*, 2014). According to the economic land rent theory, this should ideally benefit its residents, but this is not always the case as Ricardo (1889) observes. Improvement on land through physical upgrade and regularization (in form of occupancy licenses or land records) raises property values predisposing land to higher demand (Kironde, 2000; Nwanna, 2012; Thirkell,

1996). Zukin (2009) says that, when land increases in values, property owners in an area begin to sell them off to buyers who often have economic prospects of them (In Zuk *et al.*, 2015). This process organically takes place outside formal structures and is therefore referred to as informal land market (Anabaraonye, 2017). Informal land market leads to displacement of the poor from their old dwellings. This process is what Glass (1965) referred to as urban gentrification.

Among the reasons why the poor sell off their residential properties include: crisis; rise in cost of living; attractive prices; and ignorance of leasing (Thirkell, 1996; Berrisford, 2007; Kironde, 2000). The buyers often establish business enterprises while the sellers relocate to places where properties are cheaper (Nwanna, 2012; Anigacz, 2010; Tinsley, 1997). While most sellers relocate to other informal areas others rescind to the peripheries (Chikuta, 2016; Thirkell, 1996; Berrisford, 2000; Mohammed, 2014). Most scholars say that the seller's quality of life declined after the sale (Feldman, 2014; Marx; 2010; Tinsley, 1997; Thirkell, 1996; Kironde, 2000; Chikuta, 2016; Smolka, 2003; Desoto, 2001). Feldman (2014) followed up on the 10, 000 in New York City who were displaced by gentrification and found them in dilapidated houses that lacked basic necessities of life such as electricity, water, sanitation, and waste collection facilities. However, some scholars say that sellers improve their quality of life (Schill and Nathan, 1983; Vigdor, 1990; Sheppard, 2012). Vigdor (1990) who did a study in Boston found that most of the sellers are not clearly harmed by the selling but that they relocate to more descent neighbourhoods.

Studies by Thirkell (1996) and Berrisford (2007) concentrated on the comparison between who lost or won between buyers and sellers in informal land markets. The two authors focused less on whether there was improvement or decline of living conditions in the sellers' places of relocation. Rather than pursuing the sellers' outcome, their studies were about exposing the fact that the sellers were the losers while the buyers in most instances where the winners. Studies by Bruce (1993), Smolka (2001), and Nwanna (2012) were done through desk analysis. Their conclusions were based much on studies done by other scholars. They provided conclusions based on the findings of other authors. Nwanna's conclusions unlike those of Thirkell and Kironde are broad and therefore required more detailed analysis to use for comparison with the study. This study however is a field survey which collected data from the actual sellers of property. Interaction with the actual sellers of the property gave opportunity for the study to find out what their lives were before and after they sold property. The case studies by Kironde (2000), Thirkell (1996), and Tinsley (1997) considered how most of the displaced lived in squalor but did not consider their living conditions before. This study however considered the living conditions before and after the selling so as to do a comparison of life before and after the sale of properties. Findings by Schill and Nathan (1983) were made more than three decades ago, therefore, the ravages of time which often play into informal land dynamics could have changed the current state of things.

The study builds up on the work by Chikuta (2015) in Kalingalinga where he did a study on urban gentrification, morphological changes of the area and voluntary displacement. While he did not follow up on the exact places of relocation and the seller's quality of life this study pursues those ends.

1.3 METHODOLOGY

A case study approach was used because it is appropriate for in-depth understanding of a phenomenon. The sample size was 55. Eight LCC officers, two estate agents, five Kalingalinga residents -selected purposively. Thirty (30) sellers (former land owners) traced through snow ball and four buyers selected systematically. Interview guides, checklists and transcribing pad where instruments used. Stringed data was analyzed using themes, content, and descriptive statistics, whereas for numeric data a statistical paired sample t-Test was used.

The study area is Kalingalinga settlement, a place where informal land markets have been escalating ever since the area underwent physical upgrading in 1990 and was gazette as an improvement area. The map in figure 2. shows location of the study area.

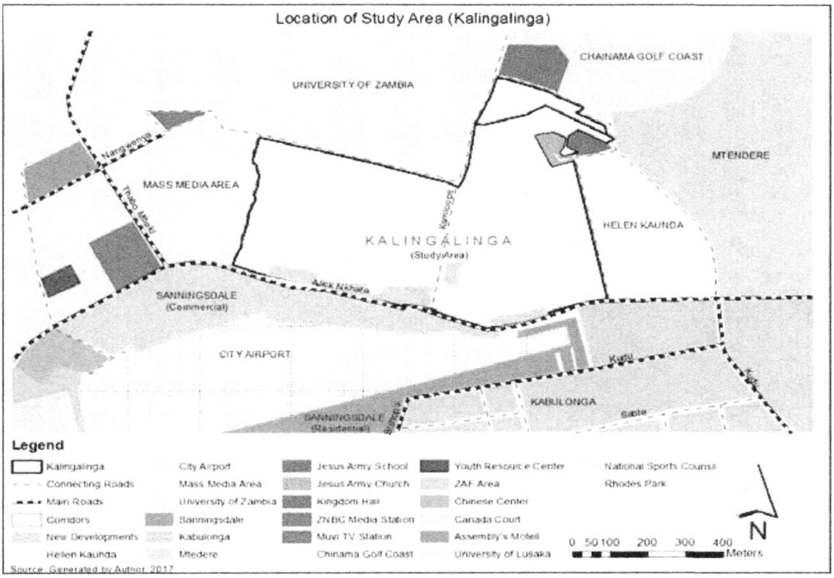

Figure 2. Showing Location of the study area

Source: Generated by Author, 2017

1.4 RESULTS AND DISCUSSIONS

1.4.1 Reasons for sell of property in Kalingalinga

The most important reasons for the sale of property in Kalingalinga are family dispute and poverty. A former land owner is quoted saying,

> *"the house belonged to the family after our parents died long ago. All of us desired to take part in the share of the property. That's the reason why we sold it. I objected to this idea of selling but my submission was overruled by majority view which anticipated higher possibilities of development after the selling"* (Pes. Com. a: 2017).

Other factors like, attractive prices, desire to improve life, and crisis also mount pressure on property owners to sell. Figure 3. shows summary of all the reasons for sale of property in Kalingalinga.

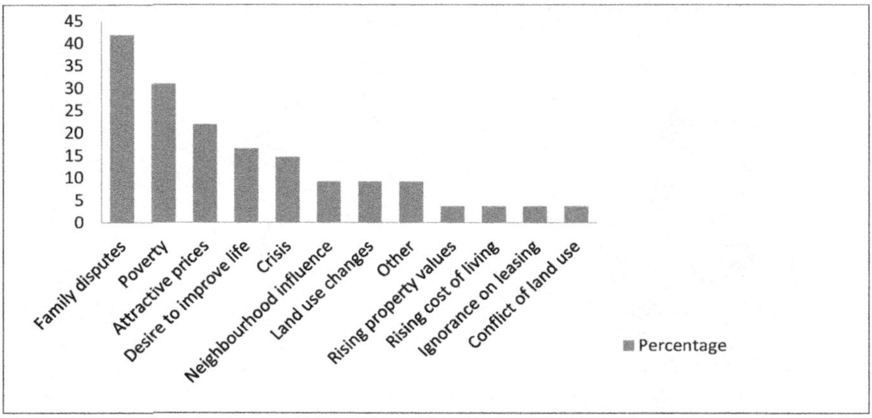

Figure 3. Showing Reasons for the sale of residential property in Kalingalinga

Source: Field Data 2017

Such findings are consistent with what Thirkell (1996) found in Cebu City, where she found that family disputes were one of the reasons for the sale of properties in informal areas. However, unlike Kalingalinga, Thirkell found that in Cebu city, what emerged as the prominent reasons for the selling were crisis and a desire to alleviate poverty among the sellers. The study is also consistent with those of Nwanna (2012) and Kironde (2000) who attributed the selling to poverty. Nwanna however did not use a case study but analyzed literature from findings in various places.

The reasons for the sale of properties in Kalingalinga are skewed towards 'desperate motives' rather than 'economic aspiration'. For example, from the findings 10 out of 12, that is, 83.3% reasons for the sell are based on a desperate situation over which the sellers have little or no option. Only two out of the 12 reasons, that is, 16.7% of the reasons were motivated by economic aspiration. This selling which is mainly motivated by desperation is consistent with the findings of Taher and Ibrahim (2014), Thirkell (1996), Tinsley (1997), and Chikuta (2016) who found that the selling of houses in informal settlements was mainly motivated by desperate situations or 'crisis' as Thirkell put it. Chikuta (2016) claimed that, although it is often anticipated that the selling is done voluntarily, in the actual sense, it is involuntary. He argues that most sellers are compelled by forces beyond them rather than a desire to improve their living conditions.

1.4.2 Places of relocation for property sellers in Kalingalinga

Mtendere East, followed by Kalingalinga, are the most important places of relocation while some sellers relocate to the outskirts of the city. Mtendere East ranks above Kalingalinga because the properties are much cheaper and security of tenure through land records is guaranteed. One seller who had relocated to Mtendere east said,

> *"they relocated to Mtendere East because land in that area was much cheaper... and this time, we have also been issued land records"* (Pes. Com. b: 2017).

Another one who relocated within Kalingalinga said,

> *"that side (further away from Alick Nkhata and Kamloops road frontages), houses are cheaper. So, in order to have remainder of money gained after the sale and therefore build another small house to rent out, we chose to relocate further away from the road frontage into the interior of Kalingalinga where land is cheaper"* (Pes. Com. c: 2017).

A key informant from LCC is quoted saying,

> *"...sellers relocate into the outskirts of Lusaka"* (Pes. Com. d: 2017).

A summary diagram in figure 4. shows the places of relocation after the sale in their descending order of significance.

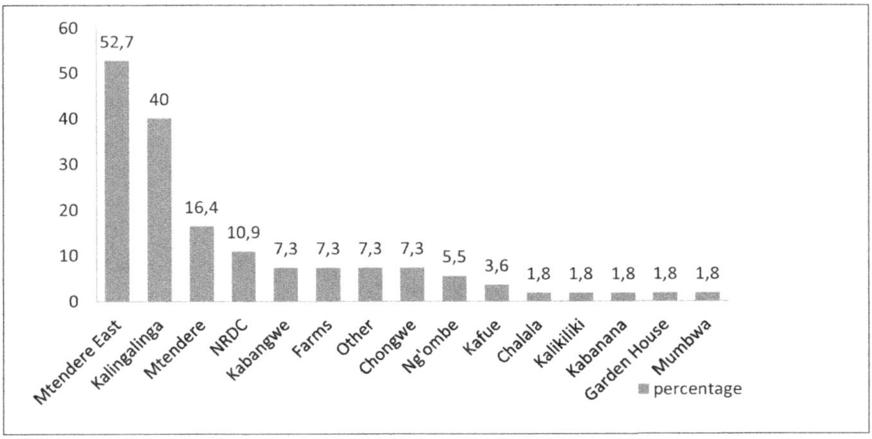

Figure 4. Sowing the Places of Relocation after sale

Source: Field Data, 2017

The findings are consistent with arguments by Kironde (2000) who said that tenure security is a pull factor which draws people to settle on a piece of land. Also, the study found what Taher and Ibrahim (2014) found in the case of India that the poor residents who are displaced often relocate to other informal areas with cheaper land. For the case of Kalingalinga, the most important reason for the relocation to Mtendere East was the affordability of the land in that area. The tenure security through the introduction of 10-year land records in the area only escalated the number of middle-income earners who relocate to the area. This is consistent with what Thirkell (1996) and Tinsley (1997) found saying that, for the most part those who relocate end up in slums because they are often without means to afford land which is either of the same or higher value as before.

According to findings of the study, most places of relocation are either informal/improvement areas or are very far away from the Central Business District (CBD) of Lusaka. Out of nine settlements in the city where the sellers relocate to, only one settlement, which is, Chalala, is the one with a background of formal designation as residential area. However, Chalala also faced challenges of basic infrastructure provision and is currently facing challenges of water supply and sewer system provision. Its residents have exploited the use of boreholes for water and are setting up private septic tanks to handle sewage. According to the planning standards, this is an unsustainable way of managing a settlement that is fast expanding (UN-Habitat, 2016). Since eight out of nine places of relocation are further away from the CBD of Lusaka, it means that 88% of the people who relocate after selling land relatively lose land that is more valuable.

1.4.3 Quality of Life for sellers in their places of relocation

There was a loss among the sellers after the selling of their properties in Kalingalinga. This loss was threefold, that is: loss in value of house; reduced access to social services and utilities; and low levels of satisfaction among the sellers after the sale. Results of this loss are summarized in the tables 1. and 2., and figures 5. and 6. Table 1. shows that the p-value indicates a significant loss in the physical value of the house. Also, table 2. shows a significant reduction in access to social services and utilities in the places of relocation. Figure 5. and 6. shows that most of the sellers who relocated regretted after selling because they did not realize the returns which they anticipated before selling.

1.4.3.1 Loss in Value of House

Findings showed that there was a significant loss in the value of houses after the selling. The p- value indicates that cost of property, number of rooms and plot sizes were significantly lower after the sale compared to the time before the sale. Only the distance to the toilet did not change after the sale as before. Since three (3) out of four (4), that is, 75% aspects of the physical household declined in value, it means that there was an overall loss in value of houses after the sale. Table 1. shows summary of the comparison in value of the houses before and after sale of properties.

Table 1. Showing Comparison in the value of the house before and after selling

Paired T-Test (CI 95%) Before and After Sale								
Valuation Variable BS (Before Selling) and AS (After Selling)	*M*	*SD*	*SEM*	*LOWER*	*UPPER*	*t*	*df*	*P-Sig (2-tailed)*
Cost of Property BS – Cost of Property AS	446034.48	172623.56	32055.39	380371.99	511696.97	13.91	29	**.000**
Plot Size BS – Plot Size AS	149.23	387.85	69.67	6.96	291.49	2.14	29	**.040**
Number of Rooms BS - Number of Rooms AS	1.03	2.31	.42	.17	1.90	2.45	29	**.021**
Distance to Toilet BS – Distance to Toilet AS	.07	1.89	.33	-.61	.76	.223	29	**.825**

Source: Generated by Author from Field Data, 2017

For the building materials the use of gravel accounted for 50% of the houses before the sale and 3.3% after the sale while the use of concrete was 3.3% before the sale and 46.7% after the sale. This indicates that there was a reduction in the use of gravel and an increased in the use of concrete after the sale of property. Since concrete is a better building material unlike gravel which is of lower quality, the findings entail that the material used for building the houses improved after the sale. In terms of toilet facility, the use of pit latrine accounted for 56.7% before sale and 26.7% after the sale, while the use of flushable toilets before sale was 33.3% and 36.7% after the sale. This entails that the use of pit latrine after the sale reduced while the use of flushable toilets increased but only by a minor percentage of only 3.4% which still meant that many people used flushable toilets even before the sell. The only improvement which can be considered more significant was the shift to using pit latrines that have sitting pans which accounted for a percentage increase of 26.7%. Summary of building materials and toilet facility before and after the sale is shown in figure 5.

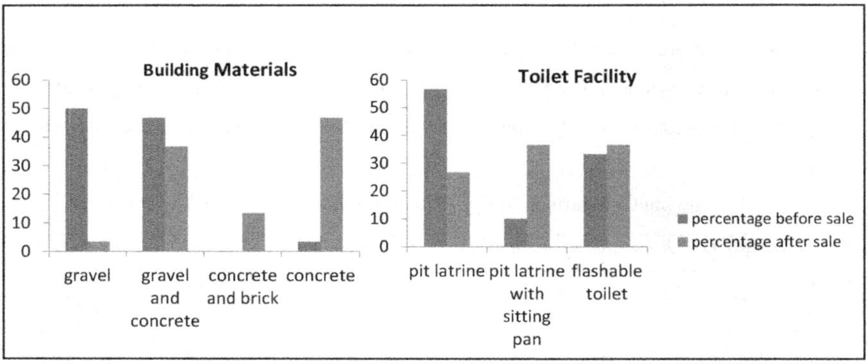

Figure 5. Building Material and Sanitation used before and after sale of property

Source: Field data, 2017

The findings show decline in the value of houses inhabited after the sale of properties. This is shown by the reduction in: cost of properties, plot sizes, number of rooms for the houses, and increase in the distances to the toilet. These changes indicate that the value of the house dropped significantly. This is similar with the findings of Tinsley (1997) in Bangkok city where he estimated that most housing units in informal settlements are of very low value. While for Kalingalinga the mean difference in cost of houses before and after the sale was as much as US$44, 603.45, Tinsley stated that particularly in Klong Toey (a place of relocation), the total cost of building a house was valued at a minimum of about US$400 which was far less that the average cost of building a descent property in any formal residential area. These findings in Bangkok are consistent with those of this study.

Although the materials used to build the houses improved from gravel to concrete, the findings also indicate that there was still a fraction of about 36.7% of those who used a mixture of both materials to build their houses after the sale. There was a tendency among those who purchased properties in Mtendere East to expand on the already existing structures made out of gravel. Tinsley (1997) indicates that household poverty in informal settlements is characterized by houses made out of building materials such as wood, gravel, and corrugated roofing sheets. This means that even though there was minimal improvement in the building material, there was still a skew towards low standard building materials because of high percentage in the use of gravel and its mixture with concrete.

The 30% decline in the use of pit latrines after the sale suggests an improvement in toilet facility but it is also important to note that 33% of the sellers before the sale used descent flushable toilets compared to the 37.3% who used flushable toilets after the sale. The only difference being only 4%. Perhaps the significant difference seemed to have been the change to pit latrine with sitting pans which accounted for a difference of 27.3% after the sale. However, this difference in the quality of toilet facility between a pit latrine with a sitting pan and one without is subjective. Some respondents felt that quality was better after they began to use the latrines with sitting pans while others felt that there was no improvement. Most sellers opined that sitting pans do not determine the quality of toilet facility unless it is accompanied by its proper use. They felt that the pans generally demand frequent cleaning for health purposes especially when the pit latrine was used by several households. Therefore, with regards value of the house before and after sale there was a decline in the quality of the sellers' livelihood.

1.4.3.2 Reduced Access to social services and utilities

The study found that there was reduced access to social services after the sale of houses. For example, the paired sample t-Test at α (0.05), 2-tailed, df (30) generated p-value of 0.000 implying a significant reduction in reduced access to health services, places of work, main market, and town centre. This reduced access to social services impacts negatively on the quality of life of the sellers in their places of relocation. The increased distance to health centers entails that many of the sellers move a very long distance to access medical care. One seller who relocated to Mtendere East said,

> *"we now have to move more than two kilometers to access a medical center, a distance we cover by walking and the journey is very tedious especially for our women. The other option of buses (public transport) and taxis (private transport) are often too costly for us"* (Pes. Com. e: 2017).

Another respondent who resettled in NRDC complained about the greater distance he has to cover to report at his place of work in Kalingalinga. He said,

> *"life is now hard, I now have to move a distance close to ten times more than the one I moved to access my place of work in Kalingalinga. Before I relocated to this place (NRDC) my place of work was literally at the backyard of my house and I had no need to cover transport costs or lose energy walking"* (Pes. Com. f: 2017).

The reduced access to town center implies more transport costs. This in turn has raised the cost of accessing commodities which most of the sellers (former land owners) sale on their retail stands. Since most of those who relocated predominantly are traders, the prices of their commodities are raised higher. This makes their goods too expensive for their local market. This creates a situation where the traders lose demand of their supply hence creating losses. A summary diagram to show comparison in terms of access to social services before and after sale is presented in Table 2.

Table 2. Comparison in the access to social services before and after selling

Paired T-Test (CI 95%) Before and After Sale								
Valuation Variable BS (Before Selling) and AS (After Selling)	*M*	*SD*	*SEM*	*LOWER*	*UPPER*	*T*	*df*	*P-Sig (2-tailed)*
Distance to Health Center BS – Distance to Health Center AS	-3.9	5.18	.931	-5.81	-2.00	-4.19	29	**.000**
Distance to Work plc BS – Distance to Work plc AS	-3.9	5.18	.931	-5.81	2.00	-4.19	29	**.000**
Distance to Main Market BS - Distance to Main Market AS	-672.58	738.34	132.61	-943.40	-401.76	-5.07	29	**.000**
Cost of Transport to Town BS – Cost of Transport to Town AS	-2.59	3.46	.621	-3.86	-1.33	-4.18	29	**.000**

Source: Generated by Author from Field Data, 2017

In terms of electricity, there was 100% access by the respondents before the sale but only 93.3% after the sale. This indicates that there was a reduction of 6.7% in terms of access to electricity after the sale. This reduction was due to relocation into areas that had no electricity connections such as Kabangwe, while some had moved to more depraved areas in Ngómbe. Since electricity is a basic necessity for life especially in an urban setting, it can be deduced that, there was a decline in quality of life due to the reduced access to this basic utility after the sale. As for access to water, there was no use of shallow wells before the sale; however, there was 6.7% of their use after the sale. There was no use of boreholes before the sale, while there was 43% use after the sale. There was 13% use of community taps before the sale and 33.3% after the sale. The use of household taps before sale was 86.7% while after the sale there was a reduction to 16.7%.

This means that, there was reduced usage of more descent sanitation. According to planning standards, household taps are the most decent access to water in any settlement (UN-Habitat, 2016). There was no use of shallow wells and boreholes before the sale of properties until after the sale. The statistics indicate that while there was increased usage of community taps there was a high reduction of 60.7% in the usage of household taps after the sale. This is because most of the sellers relocated to residences which lacked water pipes. Although the American government through the Millennium Challenge Account Zambia is installing these facilities in most of these areas, many of the residents still do not have their houses connected to the main supply pipes.

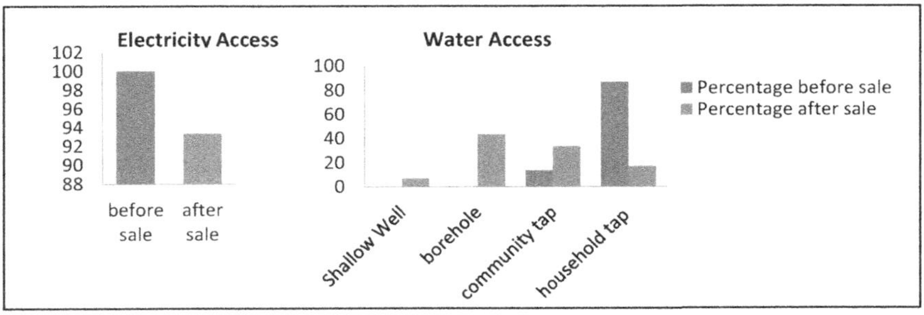

Figure 6. Showing Access to Electricity and Water before and after sale of property

Source: Field data, 2017

The increase in distance to the nearest health center, place of work, main market, and cost of transport to town center collectively indicate reduced access to social services in the areas where the sellers relocated to. The increased distance to places like town center and health centers meant that expenditure on transport to access such services increased. Since the expenditure increased after the sale, sellers were unable to access social services in the manner they did before they sold the properties. Such findings are consistent with those of Sheppard (2012) whose analysis in Metropolis found a strong correlation between displacement and increase in expenditure in the places of relocation. Using expressions in "elasticity" (beta [β] values) form to view the net effect of displacement among thousands of people, she found that several variables including the ones used in this research compounded to make costs in the places of relocation costly.

Coupled with this reduced access to social services was a reduction in access to some utilities like water and electricity. While there was 87.3% access to household tap-water before the selling, there was only 17.3% access to household tap-water after the selling. This staggering 70% reduction in accessing tap water after the sale of properties indicates a significant decline in the quality of life after the sale. According to city standards, household taps are the most convenient way to access water for domestic purposes (UN-Habitat, 2016). The findings show that most people after the selling moved from using household taps to shallow wells, and community taps. Although 47.3% of the sellers accessed water from boreholes, it is important to note that most of the sellers were not the sole owners of the boreholes but other residents within their neighborhoods. This meant that they still had to struggle to access water. These findings are consistent with what Feldman (2014) found among the displaced in their places of relocation in New York City. Among the challenges he found was their limited access to safe drinking water and poor sanitary conditions. Just as the study found 6% reduction in terms of access to electricity, Feldman (2014) also found a reduction in its access although he did not give estimates of how much. Therefore, considering this reduction in access to social services; household utilities; and increased expenditure, there was a decline in the sellers' quality of life after the sale.

1.4.3.3 Regret after selling

The extent of regret after selling is measured using the win/lose approach used by Thirkell (1996) and Berrisford (2007). While the two authors analyzed the winners and losers between buyers and sellers in their studies, this study considered the gains and losses incurred by the sellers. In this analogue, the gains are the 'win' while the losses are the 'lose'. Three standards adopted from Tinsley (1997), Thirkell (1997), and Muhammed *et al* (2014) were used to analyze these gains and losses which are: liabilities against assets; individual impression on improvement or decline; and whether 'to sell' or 'not to sell'.

In terms of liabilities incurred against assets acquired, about 43% of the sellers purchased apartments which they were apparently renting out, 8% invested in the education of their children, 6% bought vehicles, 5 % started businesses, and 38% accomplished nothing. Most sellers were renting out some sections of their properties in Kalingalinga and the 43% of the respondents who built houses for renting out after the sale lamented that, the amount of money gained from the renting was little compared to the amounts of money gained from rent before selling and relocating from Kalingalinga. For example, a wife to one of the sellers who relocated to Mtendere East said,

> *"Earnings from rentals here in Mtendere East are quite little compared to Kalingalinga. Here, for three houses, we only earn ZWM 450, while in Kalingalinga we used to earn as much as ZWM 1500"* (Pes. Com. g: 2017).

This meant that rentals in Mtendere East were threefold less than those in Kalingalinga. This decrease in amounts fetched from rentals after the sale of property meant that there was decline instead of gain even from the houses rented out. Out of the 6% who bought cars majority of them were not using them because the vehicles were either involved in accidents or lacked maintenance and were parked in their deplorable conditions. Only one of the sellers was found using his truck for beneficial purposes. A Kalingalinga resident narrated how a certain landlord in Kalingalinga sold his property and bought a car which he used for luxury purposes and eventually lost money to sustain it in the process and now has parked it. He said,

> *"The man could not believe that he held so much money at once and instead of investing it, he decided to spend it by going about womanizing, drinking beer excessively, and visiting expensive lodges and hotels. After squandering the money, he is now renting a house within Kalingalinga and surviving on piece works"* (Pes. Com. h: 2017).

The 13% who invested in education and businesses were the ones who appreciated the selling because they not only built other houses but established businesses and managed to educate some members of their households. One child to a seller confessed having been a beneficiary of the selling through the tertiary education he received in South Africa. The 38% who achieved nothing were mostly the ones who were found to be living in smaller houses which they had failed to complete. When asked about what asset they managed to acquire after the selling, most of them said that they acquired nothing apart from the unfinished houses in which they were living.

In terms of "individual feel" of improvement or decline in their quality of life, majority (45%) of the respondents indicated that the lives of most sellers declined. About 23% had mixed impressions about whether or not to sell given a second chance, indicating that their lives both declined and improved in certain ways. However, going by the percentage distribution of 45% improvement and 23% mixed view, it was clear that most sellers lost out. Most of the sellers lost 'prime land' in Kalingalinga even though some of them acquired other unfinished properties elsewhere which they were struggling to complete due to limited resources.

When asked for an opinion whether they would sale their property or not given their situation before selling, majority (69%) of the respondents opined not to sell the properties in Kalingalinga, 11% were not sure, while 20% opined to sell. From this percentage distribution again, majority of respondents opined not to sell. A ratio of 69% to 20% for the 'not to sale' versus 'to sale' shows how much the sellers regret having sold their properties. On average many sellers had higher expectations from the selling and were in one way or another frustrated later on by the whole undertaking. In expressing regret, one seller who moved to NRDC area lamented saying,

> *"Sir, that topic of selling properties is sensitive to me and grieves my heart. Whenever I pass by the place of my former house in Kalingalinga along Alick Nkhata road, I*

always wish there was a way of reclaiming my land so that I would never sell it again" (Pes. Com. i: 2017).

This statement of despair shows how the sellers deeply regret the act of selling their properties. Figure 7. shows a summary of the sellers' view after selling, assets and liabilities accrued after sale and 'individual feel' of satisfaction or regret after the sale.

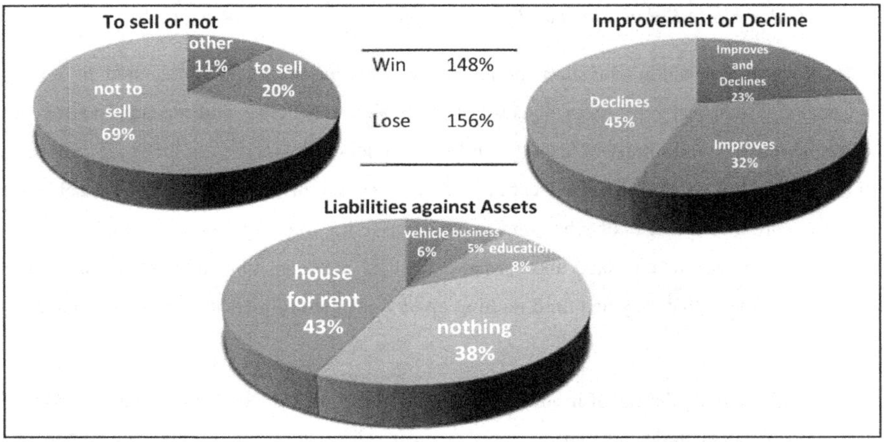

Figure 7. Showing the percentage ratio of the win-lose situation of property sellers

Source: Field data, 2017

Most of the sellers who had relocated regretted having sold off their properties. The reduced earnings from rentals; incurring of liabilities instead of assets; and 38% of the respondents' achievement of nothing, indicates a general loss of "prime land" and valuable properties. Such findings which indicate loss are consistent with those of Thirkell (1996) who stated that most of the poor whose property is bought off by the richer middle class lost out more valuable land which they only realize later. Smolka (2003) also says that, the value of land which the rich buy from the poor in the inner city is often higher than what its occupants (poor owners) anticipate, and so they sell in ignorance and therefore are often at a loss.

The 68% who opined not to sell given an option before the selling, show contrasting findings to those of Twarambamenye and Nyawidi (2012) in Butare town, where they found that, "the majority of respondents (55.9 %) consider that informal land transaction benefits either the seller or and the buyer separately or both." It is however worth noting that their findings do not clearly separate the ones who really benefits separate between buyers and sellers. This study has streamlined that in the case of Kalingalinga the sellers lose out.

1.4.4 Spatial implications of informal land markets

These informal land markets in Kalingalinga result in construction of large buildings of modern standard on spaces where the land is sold in Kalingalinga and inferior infrastructures on spaces where the land is bought in the places where the sellers relocate. For the structures on the spaces where property was sold off in Kalingalinga, buyers have built multi-storey buildings and are trading in all kinds of goods and services. One shopping mall owned by some Korean nationals, has seen a fill of lots of people purchasing goods and services within a space of less than three months before its completion. The scene in figure 1.6, pic 1 depicts an example of what one will see from many other places where land was purchased in Kaingalinga by the investors. Clearly for the buyers, informal land markets mean profit-making as Thirkell (1996) and Berrisford (2007) found, and Marx (2010) said, that, the purchasing of properties in informal settlements by the richer middle class in a society is motivated by economic prospects other than residential.

Contrary to the multi-storey buildings which one finds at the land where land was sold off in Kalingalinga, the images on the places of relocation have structures of inferior standard compared to those on the land that was sold. According to interviews with the local estate agents, the structures shown in figure 8., pic 1 are sited on the land where most people who lived on the space in figure1.6, pic 2,3, and 4 lived before they sold. Picture 4 shows one of the houses belonging to a seller who built it using low quality materials. Picture 2 is an unfinished toilet infrastructure. The photo number 3 is an unfinished apartment intended to be rented out. A summary for the comparison between the structures built on spaces after the sale and on spaces in the places of relocation is shown in figure 8.

Figure 8. Showing comparison of structures on spaces before and after sale of property

Source: Field data, 2017

1.5 RECOMMENDATIONS

The study recommends that a framework to guide informal land markets in Kalingalinga be put in place so that the plight of the poor sellers is taken care of. There is also need for evaluation of prices for selling the residential properties so that the poor are able to leverage themselves in their places of relocation and achieve what Marx (2010) proposed when he said that, "if informal land markets are well natured by formal institutions, they can be a source of revenue that can bring assets to the sellers and redeem them from their poverty situation". There is need to put in place policy which will ensure preparation of inclusive upgrading plans to achieve inclusiveness in urban planning since the world is on a trajectory to achieving SDG number 11 in the terms of spatial planning. The Lusaka City Council may consider preparing a plan to ensure a smooth transition of Kalingalinga from a residential to a commercial area so that there is no conflict of land use between residential and commercial. This conflict of land use implies conflict of interest especially that the investments are of serious implication in the neighbourhood. The study also recommends for further research to be done in order to establish how formal structures can set up policy which can realign informal land markets to protect the interest of the poor sellers.

1.6 CONCLUSION

Most people sell residential properties due to reasons that are motivated by desperation and the lack of choice. They relocate to places with backgrounds of informality and far-away places which create costs for them. Their property values after selling reduce and much of their livelihoods decline. Most sellers regret and wish they could reclaim the land which they sold in Kalingalinga. The implications of informal land markets on the informal sellers of land in Kalingalinga are mostly negative. The disadvantages of these land markets are found in: the desperate reasons why the residents sell their properties; their depraved places of relocation after the sale; and decline in their quality of life after the sale. It is therefore true that urban gentrification in African cities can be a menace when the physical upgrading process is not coupled with socio-economic upgrading. This is because as observed in Kalingalinga, the physical upgrading has predisposed the area to investors who have taken advantage of informal land markets to dispossess the poor residents of their land. Therefore, there is dire need for teams which do the upgrading of informal settlements to consider the implications of such dynamics if those upgrading processes will be sustainable and meet the Urban Challenge of 2014 and the SDG number 11.

REFERENCES

[1] Anabaraonye, S. O., 2017. Determinants of Informal Land Transactions on Land Market in Owerri Urban, Imo State, Nigeria, *Civil and Environmental Research., 9 (9), 17, 70-76.*

[2] Anigacz, A. A., 2010. *Are Residents Who Are Displaced by Gentrification Better or Worse off After Relocating?* Honor Projects, Paper 28, Macalester College. http://digitalcommons.macalester.edu/economics_honors_projects/28

[3] Berrisford, S., 2007. *Do Informal Land Markets work for Poor People? An Assessment of Three Metropolitan Cities in South Africa: Research Method and Survey Design.* Isandla Institute, Kenilworth.

[4] Chibuye, M., 2014. Interrogating Urban Poverty lines: The case of Zambia. *Environment and Urbanization, International Institute for Research and Development., 26 (1), 236-256.*

[5] Chikuta, E., 2016. *Implications of Informal Settlement Upgrading on Urban Development in the City of Lusaka: The Case of Kalingalinga.* A paper presented at the Urban Research Forum Maputo, Mozambique on 7th June 2016, 4-7.

[6] Tinsley E., 1997. *Mechanics of Informal Land and Markets, A theoretical Position.* Dissertation: Massachusetts Institute of Technology, 50-61.

[7] Feldman, J., 2014. *Gentrification, Urban Displacement and Affordable Housing: Overview and Research Round-up.* Harvard Kennedy School: Cambridge.

[8] Giannias, D. A., 1996. *Quality of Life in Southern Ontario.* Department of Economics, University of Crete, Athens Greece.

[9] Kironde, L. J. M., 2000. Understanding Land Markets in African Urban Areas; the Case of Dar es Salaam, Tanzania. *Habitat International., 24, 151-165.*

[10] Macpherson, L., 2013. Participatory Approaches to Slum Upgrading and Poverty Reduction in African Cities. *Hydra Interdisciplinary Journal of Social Sciences.*

[11] Muhammed E., S. M. Ali, and U. A. Nasar. 2014. Economic Land use Theory and Land Value Model. *International Journal of Economics and Statistics., 2, 1-2, 93.*

[12] Nwanna, C. R., 2012. Gentrification in Lagos State: Challenges and Prospects. University of Lagos, Lagos, *British Journal of Arts and Social Sciences., 5 (2), 165.* British Journal Publishing.

[13] Sheppard, S., 2012. *Why is Gentrification a Problem? Center for Creative Development.* Williams College: Williamstown, 1-3.

[14] Smolka M. O., 2003. *Informality, Urban Poverty and Land Market Prices. Dissertation, Cambridg.* Lincoln Institute of Land Policy Research Paper, 7.

[15] Taher, T. M. and A. Ibrahim. 2014. Transformation of Slum and Squatter Settlements: A Way of Sustainable Living in Context of 21st Century Cities. *American Journal of Civil Engineering and Architecture., 2, (2), 70-76.*

[14] Thirkell, A. J., 1996. Players in urban informal land markets; who wins? Who loses? A case study of Cebu City. *Journal on Environment and Urbanization., 8, (2), 78-86.*

[15] UNDP 2014. *Human Development Report: Sustaining Human Progress: Reducing Vulnerabilities and Building Resilience.* Published for the United Nations Development Program.

[16] UN-Habitat 2016. *Strategic Development Goal Number 11 Key Indicators, Population and Development Branch*: Geneva.

[17] UN-Habitat 2011. *Zambia Participatory Slum Upgrading Program (PSUP).* Funded by the European Commission.

[18] UN-Habitat 2010. Urban land market: economic concepts and tools for engaging in Africa, Retrieved on 20th June 2016 from www.unhabitat.org.

[19] Zuk, M., A. H. Bierbaum., K. Chapple., K. Gorska., A. Loukaitou-Sideris., P. Ong., T. Thomas, 2015. *Gentrification, Displacement and the Role of Public Investment: A Literature Review.* University of California: Los Angeles, 15-31.

CHAPTER TWO: DYNAMICS IN THE HISTORY AND ECONOMIC DEVELOPMENT OF MAN; REFOCUSING ON ECOLOGICAL DISTURBANCE AND CLIMATE CHANGE

ABSTRACT

Man's history and development endeavours have been advancing alongside a trail of ecological ramifications and climate change. Since prehistoric times, scientists have not recorded an accelerated shift in the ecology of the planet during any other epoch beside that of modern man. The paper seeks to explore how man's history and development affects ecology and climate. It uses desk analysis to recollect data from global assessment reports and runs a One paired Sample Means t-Test, 1 tailed, 8 df, at Pearson Correlation value 0.458 and 0.5 alpha level. Findings show that, there is global climate change, seen in global warming trends; and imbalance in ecological footprint, seen in depletion of air, water and land sinks. The t-Test reveals significant net loss of global forest cover. The study also, apparently found that, processes of development generally tend to damage ecology. Therefore, the study recommends a refocus to sustainable means of development.

KEYWORDS: *Ecological Footprint, Climate Change, Sustainable Development, Built Environmental, Integrated Planning*

2.1 BACKGROUND AND INTRODUCTION

From age to age, the evolution of man has been proliferated by significant stages of development. According to mainstream history, man has long been advancing himself to find better means of meeting his needs (Arlt and Daviau, 2009). Unlike his earlier days on the planet, man's motivation for development is now driven by a plethora of multiple contingencies (Wengrow, 2018). From the pre-Neolithic age when the urge to develop was primarily driven by survival instinct, today such motivation is also driven by pleasure, luxury, pride, and superiority (Hale, Ricotta, Freed, Smith, and Huang, 2019). His evolution followed a development path from the early, middle and late stone age; to the early and late iron age (Muhanna, 2017). From there, it advanced to the industrial revolution and the Age of computers or Artificial Intelligence (AI) (Ibid, 2017). Evolution significantly altered the way mankind works, produces, packages, trades, consumes and disposes, causing negative impact on ecology. The study endeavours to show how compromised mode of production, consumption and waste disposal projects on the future of the planet. It systematically analysis how history can teach us a lesson about how mismanagement of natural resource can jeopardize ecology. It also plots how human development has been advancing alongside the speedy loss of ecological footprint and exacerbating climate change.

The first part of the paper defines the scope of the study area and outlines the problem statement; aim; objectives; hypotheses; research questions and rationale of the study. The second, defines the key concepts and reviews their interplay of each other. The third part outlines the methodology of the study indicating the target population, revealing sampling technique and outlining collected collection/ analysis. The fourth chapter presents and discusses the findings of the study. The fifth chapter draws the conclusions and makes recommendations about the findings of the study.

2.1.1 Scope of the study area

The study has taken a global perspective of climate change because, first, the phenomenon can be established when the multiplicity of variables defining it are gathered across a relatively wide geographic scale (Harvey, 2018). Second, the matter of climate change is of global implication. The analysis selects some parts of the world where there are indicators pointing to climate change.

2.2.2 Statement of the problem

Human evolution has seen the development of man from one age to another. Civilization with its unyielding quest to make life easier made operations easier and speed up processes in raw material extraction, transport, communication, manufacturing, processing, trading, and the consumption of goods and services as well as waste disposal. The industrial revolution of Europe in the late 1700s came with mass production and with it, mass generation of waste (Clark, 2014). The unprecedented effects were hazardous -pollution, environmental degradation and climate change. The consequences of these hazards are unforgiving -global warming, disturbance in weather patterns, loss of catchment, flooding, heightened frequency of tropical storms, loss of fresh air and water, outbreak of epidemics and disease, overcrowding in urban centers, sprawl, urban gentrification and depletion of green spaces (Rymarczyk, 2020). The study seeks to highlight this predicament in the light of man's evolution. Although some studies have been done to show trends in climate change and its ramifications, their still remains much to be exploited about history's direct contribution to the matter of climate change and environmental degradation.

2.2.2 Aim

The aim of the study is to establish the extent to which human development has contributed to climate change and the erosion of ecological footprint over the past 500 plus years, and show how all this has affected quality of life on earth.

2.2.3 Objectives

i. To explore the dispensations of human civilization over the past five plus centuries
ii. To determine the extent of climate change over the past five plus centuries
iii. To establish the correlation between human development and climate change
iv. To assess the net change in global vegetative cover in a 67 year-period from 1948-2015
v. To ascertain the ecological ramifications of climate change on wildlife and the quality of human life

2.2.4 Research Questions

Objective (i)

What are the most important stages of human development in the past 500 years?

Objective (ii)

Has there been climate change on earth within the past 500 years of human history?

Objective (iii)

Is there a correlation between human historical development and climate change?

Objective (iv)

What impact has climate change had upon the ecology, biodiversity and quality of human life?

Hypotheses

Objective (v)

H_0: There has not been global net loss in forest cover over built environment during the 67-year period from 1948-2015.

H_1: There has been global net loss in forest cover over built environment during the 67-year period from 1948-2015.

2.2.5 Significance of the study

The study endeavours to surface information that can contribute to the debate around human development versus climate change. It assembles data that shows a clearer perspective about the way human history interweaves into the fabric of ecological transformation and climate change. Such knowledge can bridge the gap in literature on how human evolution plays a role in the path of sustainable development. Also, the study can provide global institutions, agencies, and NGOs with relevant data to help them prepare world trajectories against climate change. Apart from that, the information gathered can furnish civil engineers in environmental agencies with grassroot indicators for risk assessment, and to build instruments for measuring ecological depletion. This way, they can better inform policy. It can also serve as a guide for integrated planning among urban planners. The historical background of man's development provides warning signs against impending doom to communities whose industrial advancement is still young and vibrant, such as in African and generally -the global south, so that they can avoid mistakes that more advanced civilizations made in the past.

2.3 LITERATURE REVIEW

2.3.1 The historical development of man and technology

The history of man over the past 500 years can generally be demarcated into the Neolithic Age, the Iron Age, the Age of Industrial revolution (Knowledge Explosion), and the Age of Computers (Artificial Intelligence) (Scarre and Fagan, 2016). During the Neolithic period, man's way of life was simple, with an economy mainly consisting of hunting and gathering on small-scale primitive subsistent farms (Ibid, 2016). Before BC2000, communities in many parts of the world except for the Middle East were primarily hunter-gatherers (Muhanna, 2017). Agriculture had not yet gained ground and settlement infrastructure was basically made up of small grass-thatched houses made out of mud (Arlt and Daviau, 2009). In the progress of time, populations grew and the Industrial Revolution began (Clark, 2014). Sources indicate that, it started in Britain during the 1700s following the explosion of knowledge in science and technology (Xu, David, and Kim, 2018). With the invention of a powered engine, it was now possible to harvest and process raw materials in larger quantities (Ibid, 2018). Countries across Europe and overseas began to industrialize and this came urbanization (Rymarczyk, 2020). According to Rymarczyk, urbanization came with pressure on settlement because, people needed constant water and food supply, shelter, education and health facilities, recreation and waste management systems.

From the foregoing, it seems undeniable that the industrial revolution is one of the most important transition of man's development and appears to make the greatest impact on environmental degradation. This is because it came with mass deforestation, draining of marshes, emission of chlorofluorocarbons (CFCs), and discharge of effluence (Rymarczyk, 2020). All these resulted in the erosion of green spaces, the loss of biodiversity, drying up of surface water bodies, lowering of the water table, and pollution in all its forms (Rashid, Irum, Malik, Ashraf, Rongqiong, Liu, Ullah, Ali and Yousaf, 2018; Jorgenson, 2003). The industrial age has also seen another advancement in the development of man called the Computer Age. This heightened use of machine technology in almost all economic sectors has truncated another pressing demand for energy (Rymarczyk, 2020). The correlation between economic growth and environmental degradation seems to puzzle even the elite of elites. The dilemma is to reconcile the gap between advanced communities and those lagging behind by strengthening economies while preserving ecological footprint for posterity.

2.3.2 Climate Change

According to Harvey (2018), climate change is the significant change in the overall pattern of weather defying the long-term pre-existing or natural conditions. Climate prevails over a long time and rarely changes. Suffice to note that, geological evidence shows that climate has been changing from epoch to epoch (Singer, 2006). The reasons for such changes are still being researched. Nevertheless, it is assumed that the modern human race (homo sapiens) and their associated technology have been on the face of the earth for less than 20 000 years, and 300 years ago climate change seems to have taken rapid shifts (Labat, Goddéris, Probst and Guyot, 2004). More robust assessment to unearth the way climate was changing in prehistoric times is being done to establish whether our contribution to the predicament is significant or not. Nonetheless, climate change itself is a part of a malfunctional ecosystem attempting to balance itself (Held and Soden, 2006). This indicates careless manipulation of both the living and non-living environment which are the sources of natural resources. Such things could have happened in the past, perhaps as a function of natural terrestrial or extra-terrestrial catastrophe. However, in man's history, the age of industry has its fair share of contributing to the predicament. In the meantime, it is phronesis to act prudently and create a preventive strategy against future calamity. The first step is to understand the dynamics involved, and for the purpose of the study a historical exegesis of how human civilization once mishandled can disturb natural ecology has been used to establish such knowledge.

2.3.3 Ecological Footprint

The earth is a unit mass of matter found in the dynamic complex of nature governed by universal laws. It is made up of an environment that functions by a symbiotic balance of living and non-living organisms. Ecology can basically be described as the interaction between the living and non-living environment as well as their sub-interaction to create natural balance and continuity (Rees, 2018). Ecology is a system that constitutes habitat and symbiotic relationships among organisms. Its footprint spans across the entire earth with an earth-atmosphere system is sustained by a complex network of interaction between endowments on the earth's surface and atmospheric forces (Held, and Soden, 2006). For instance, if vegetative cover is sufficiently eroded, then river catchment is lost and the rate of evapotranspiration is reduced (Labat, *et al.*, 2004). This in turn reduces the amount of rainfall received in a catchment. If such a situation is extrapolated across a broader space, then it significantly transforms the average conditions for smooth interaction in the earth-atmosphere system. Such global ramifications are due to erosion of the ecological footprint, which in this case can be defined as, the removal of original/natural resources (vegetative cover) of the earth.

The world in on a trajectory to find means of striking a balance to cope with the 'poverty-environmental degradation' trap. The most significant strides are seen in the advancement and implementation of the SDGs (UN-Habitat, 2019). This new trajectory has given birth to fresh concepts and methods in the field of planning, such as regional integration, inclusivity, mixed land-use, green building, garden city, and 'city within a city' (UN-Habitat, 2022). Therefore, planners have restructured the concept of planning to suit the growing need to preserve ecology. And so, land-use planning is reshaping to ensure integrated mixture of both natural and built environment (Dovey, Shafique, Van Ostrum and Chatterjee, 2021; Ibid, 2022). The study reveals the conflict between urbanization and ecology preservation.

2.3.4 Development and Climate Change

Development refers to change of something, in this context, improvement or progress (Klasen, 2018). Man's development implies advancement in the modus operandi to create capacity for meeting all of a population's basic needs (Ibid, 2018). This definition rests upon the bedrock of optimal and balance of population. Optimal population refers to the sufficiency of a population to maximize its natural resource for its benefit, while balanced population refers to the ability of natural resources to sufficiently furnish the population size (Yumashev, Ślusarczyk, Kondrashev, and Mikhaylov, 2020; Rabie, 2013). While the ratio between population size to resource base is critical, it is the ability of the population to access basic needs of life that really matters. This involves the transformation of raw materials into products and services and avail them to the mass of the people (Yumashev *et al.*, 2020; Klasen, 2018). The process of converting raw materials into finished goods often requires industry. The larger the population, the higher the demand and consequently the large the industries. If production is deficient, then goods and services become scarce which leads to inflation. Therefore, a healthy Gross Domestic Product (GDP) is key for the development of a country (Klasen, 2018).

Also, the mass of the people should have access to these resources which would translate into favourable Human Development Index (HDI). The index refers to income per capita or the amount of money that an individual in a country will have to spend for an entire year if all the resources of the country were equally distributed (Ibid, 2018). This entails that, healthy economies inherently have individuals within a population sustaining enough buying power to decently meet all their basic needs and more (Chibuye, 2014). Nonetheless, economic growth must not override resource repletion and waste management. Otherwise, a growing economy that lacks resources to sustain it, alongside a contaminated environment compromises quality of life by brewing depravity, poverty, sickness and disease, hunger and starvation, and even death. For civilizations that have continued on the trajectory of economic development negating environmental concerns, the ravages of climate change have come with unforgiving consequences.

2.3.5 Climate Change and its ramifications on quality of life

Climate change has a telling impact upon the quality of both human and non-human life. The most obvious effects of climate change are witnessed in extreme weather, heat waves, tropical storms, torrential rains, floods, droughts, increased glaciation, and global warming (Bartke, Forster, Githiri, Jering, Kago, Schlimmer, and Sietchiping, 2021). These outcomes of climate change have both direct or immediate repercussions as well as indirect and long-range repercussions, impacting the quality of human life on earth. Quality of human life can be defined as the measure of access by a person to basic needs (Chibuye, 2014). The most basic of such needs is safety and security, food and nutrition, fresh water and air, health and descent shelter (UN-Habitat, 2018). Therefore, it always remains crucial to provide an environment that sustains the provision of these things. Hidden from the radar of economists and most developers are the far-reaching damages caused by climate change. For example, climate change can cause heat waves that result in tropical storms leading to death (Pieterse, 2010). Meanwhile, the ripple effects of tropical storms or tsunamis are the floods that sweep across houses causing water pollution and consequently water-borne diseases (Nchito, 2007). Also, the torn down infrastructure sometimes leaves irreparable damage which bites into the economy, while the loss of lives leaves emotional scars (Vlek and Steg, 2007). Although we may perceive it to be a simple matter, the underlying reality is that, what really enables any organism to exist is its host (environment) (Ndebele-Murisa, Mubaya, Pretorius, Mamombe, Iipinge, Nchito, Mfune, Siame, and Mwalukanga, 2020).

2.4 METHODOLOGY

2.4.1 Research Design and Approach

The study is a desk analysis that surveyed data from a relatively large number of places in the world. Since the study sought to cover a wider geography, the desk analysis was more ideal (Bostley, 2019). The study recollected data from remote sensing surveys done by the Forest Resource Assessment (FRA) (FRA, 2020). These were extracted from photointerpretation of a global sample of high-resolution imagery through a worldwide network of national experts. The surveys involved the use of modern remote sensing tools and methods. The study approach was a mixed method. Quantitatively, the study recollected large quantities of geospatial images, and therefore used the count of hexagonal pixels obtained from Landsat and Sentinel Mosaics. The recollected geospatial images were captured and measured by surveyors from the National Research Council (NRC) and the areal cover of ecology and trends in climate change was recorded year after year (NRC: 2015. Data on forest cover, was analysed using statistical testing. Meanwhile, qualitatively, the study explored data recollected from field surveys done by the FRA via interviews of indigenous people in several countries across the globe.

2.4.2 Sampling Technique

The sample is drawn from a population of spaces mapped out of a global grid of equal area hexagons documented by the FRA. A stratified random sampling technique was used by the agency to conduct the survey and reduce uncertainty of the forest area change estimates. The technique was based on eighty (80) strata derived using a combination of twenty (20) Global Ecological Zones for categories of change in the cover of trees. The map under assessment had a total of about 430 000 remote sensing samples of hexagons covering 40 hectares and centroid spreading across 1 hectare from all over the world. Purposive sampling was used by the FRA to select indigenous people for interviews across several countries. Also, countries were selected based on evident disproportionate removal of forest cover or widespread built environment. Apart from these, the study also purposively sampled timeline analysis graphs and images from the NRC showing global trends in the change of climate. The graphs were selected based on the climatic index they were projecting and its significant contribution to the nature of climate and ecology. In this wise temperature, land, ocean, air, and vegetation sinks were part of the variables.

2.4.3 Data collection instrument

The data was collected through carefully reading the Global Forest Resource Assessment report prepared by the FRA from the early 1700s upto the year 2020. The assessment was done from the data obtained from remote sensing and survey. The FRA and NRC conducted remote sensing through the Collect Online Earth Platform, and Google Earth alongside Bing Maps to collect auxiliary data. Other modern geospatial tools with improved data entry, data visualization, review and analysis functions were also used to collect data, especially from countries where forest information was limited or not available. The surveys were done via interview of local residents who were familiar with the geography of the selected countries. Finally, some pictures were also recollected from the assessment reports and press releases.

2.4.4 Data Analysis

The study analysed data from the global assessment reports on change of global forest cover over the years using a One paired Two Sample Mean t-Test at Pearson correlation 0.458, 1 tailed at 8 degrees of freedom and alpha level 0.5. The other data was interpreted from graphs showing trends of climate change over the years extrapolated from the global data assessment of greenhouse gases, temperature, forests, land and water sinks. These were compiled by the NRC using the Open Floris Collect Earth software tool to visually interpret about 213 000 sample plots. The Collect Earth tool was also used in the analysis of data to access and display high resolution images from Google Earth and Bing Maps. The assessment tool also generated grids through Google Earth Engine, that allowed ready access to Landsat and Modis image archives.

2.5 FINDINGS OF THE STUDY AND DISCUSSION

2.5.1 Pre-Historic Times and Climate Change 800, 000 years ago: The Ice Age

Climate change is not a recent phenomenon. In the past epochs, the earth underwent various phases of climate change. One of the most recent traceable epochs of climate change is the Miocene epoch within which is encapsulated in the Ice Age (Rashid *et al.*, 2018). During this epoch when modern human life had not yet evolved, the earth underwent climate change when global temperatures dropped below the 'normal' average. Temperatures drastically dropped all across the world and this phenomenon resulted in the freezing of almost all parts of the earth ending most life forms (Rashid *et al.*, 2018). One theory explaining this was the collision of a meteorite into the earth causing ecological catastrophe (Rashid *et al.*, 2018; Rees, 2018). The other theory, advanced by some geological studies postulate that, the earth naturally undergoes periodic rise and fall in global temperatures (Rees, 2018).

The graph in figure 9 shows the trend of temperature during the Ice Age. One important observation made by the NRC on climate change during Ice Age and the post-modern climate change is the vast difference in temperature variation between them. During the Ice Age, after BC800 000 to around BC100 000, Carbon dioxide (CO_2) and temperature levels in the atmosphere underwent cycles of change every 100 000 years. Each 100 000 years division saw a rise and decline in temperature levels systematically. As the CO_2 levels rose, so did the temperature and the vice versa. The cyclical pattern shows a system of balance in the climatic patterns separated by a 200 000 years interval. For instance, CO_2 levels rose only upto a maximum threshold of 300 parts per billion (ppb) during the entire ice age period and this was the time when temperatures also rose to a maximum average threshold of about +4°C. However, after the Ice Age, CO_2 alongside temperature in the atmosphere have been increasing. As at 2010, the amount of CO_2 in the atmosphere was over 390ppb varying from the maximum overage of the Ice Age by a staggering 90ppb. Compounded to this accumulation of excess CO_2 in the atmosphere is the further addition of other heat retaining gases (CFCs) in the atmosphere such as Methane (CH_4) and Nitrous Oxide (N_2O), which together with carbon dioxide are accelerating the rise of global temperature.

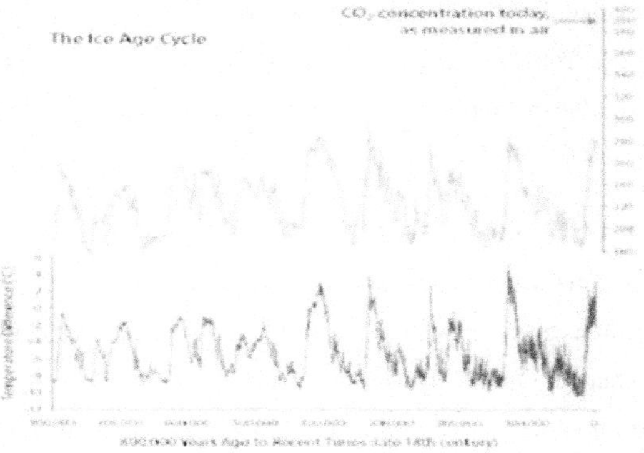

Figure 9. Showing Climate Change during the Ice Age

Source: National Research Council, 2010d

2.5.2 Important dispensations of human civilization over the past five centuries

From pre-historic times, human civilization has seen the development of technology from the simple to the complex. The human economy has advanced from hunting and gathering to agriculture, from agriculture to iron-working, and from Iron-working to the period of knowledge explosion which ushered in the industrial revolution in the 1700s. Invention during the industrial revolution gave way to the computer age or the age of artificial intelligence (AI). This series of indelible advancement in human civilization is shown in table 3.

Table 3. Showing Historical times and human civilization after the Ice Age

S/N	*Years of Invention*	*Age (Dispensation)*	*Time-line (Period)*	*Human Development (Civilization)*
1	> BC10000	Prehistoric	> 10, 000	Hunting & Gathering
2	BC10000s	Neolithic	10, 000 – 2, 000	Agriculture & Settlement
3	BC2000s	Iron Age	2, 000 – 500	Technology
4	AD1700s	Industrial Revolution	500 – 200	Manufacturing & Urbanization
5	AD1900s	Computer Age	200 - date	Artificial Intelligence

Source: Adapted by Author, 2022

These dispensations of human civilization have not gone without impacting the trends of global climatic shift. It may be true that the natural tendency of the earth is to undergo changes in weather patterns after a certain period. However, the rate at which the changes are happening in recent times is alarming. The phenomenon also reveals a seeming correlation between human civilization and climate change.

2.5.3 Extent of Climate Change during the past three centuries

Findings of the study have also shown that, the reality of climate change has been radically observed in the past three centuries beginning from the late 1800s to the year 2000. During this period, there has been a phenomenal rise in global land and ocean temperatures. Temperature indices on both the land and continental plains have shown a rise such as never been recorded since pre-historic times. The sharp rise in temperature levels shows a net rise of a +6 score within only a space of approximately three centuries. The figure 10 shows the trends in the sharp rise of temperature within the 200 plus years.

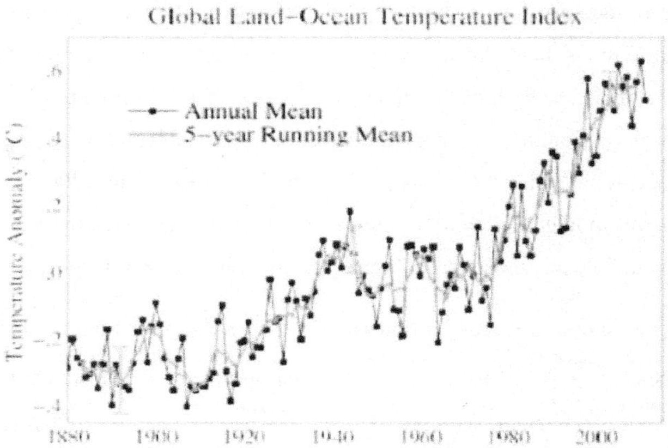

Figure 10. Showing the rise in global temperature in the past three centuries

Source: https://nap.nationalacademies.org/read/14673/chapter/2#3

There is still more evidence of global climate change in the world today. For instance, the past two centuries have seen a rise in the atmospheric concentration of Carbon Dioxide (CO_2). According to the earth system research, there has been a sharp and steady rise in the concentration of CO_2 in the earth's atmosphere. This means that, CO_2 has continued to increase steadily in the atmosphere since the 1960s as shown in the figure 11.

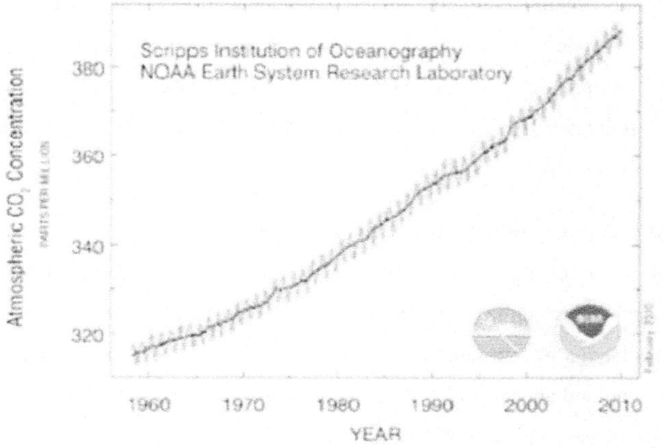

Figure 11. Showing a global rise in atmospheric CO_2 concentration

Source: National Research Council, 2011a

It is not only the rise in CO_2 concentration in the atmosphere but, findings also show that alongside are two other Chlorofluorocarbon gases that have also sharply risen. These are Methane (CH_4) and Nitrous Oxide (N_2O) which, just like CO_2 are gases with High Heat Retention Capacity (HHRC). The diagram in figure 12 shows the trend of the three gases on the atmosphere for the past 2000 years.

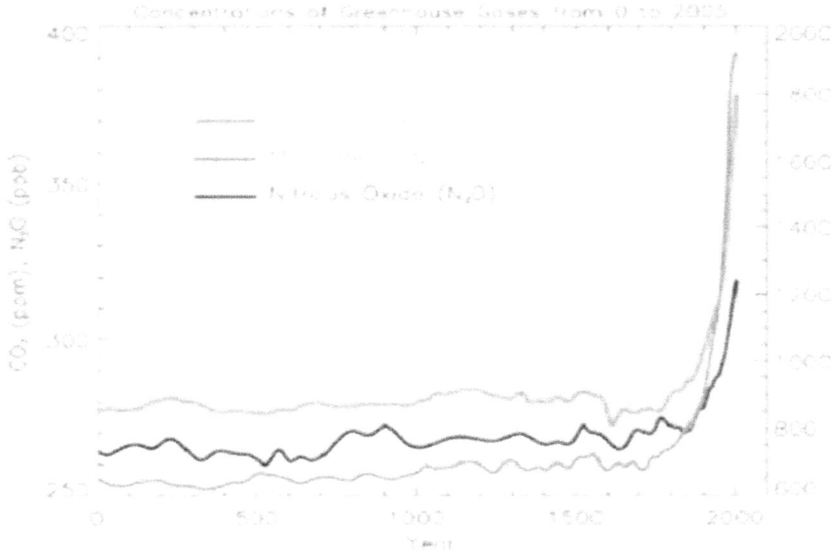

Figure 12. Showing a global rise in atmospheric CO_2, CH_4 and N_2O Concentration from the late 1800s

Source: National Research Council, 2011a

According to the figure 6, during the past 2000 years in the period from 0AD upto the 1800s, the concentration of the three gases -CO_2, CH_4, and N_2O was relatively the same. Findings show that, there was no significant rise. The lab tests reveal that, within the period from 0AD to 1800AD, on average CO_2 only had a concentration range variation of about 15parts per million (ppm), CH_4 had a concentration range variation of only about 250ppb, while N_2O had a concentration range variation of only about 15ppb. However, the graph shows that immediately after the late 1800s, all three gases shot up to levels never seen before. For example, CO_2 shot up from its normal average of 280ppm in the year 1880 to a staggering

380ppm in the year 2000, N_2O shot up from 270ppb in 1880 to about 320ppb in the year 2000, meanwhile CH_4 shot up from an average of 800ppb in 1880 to a whooping 1950ppm in the year 2000.

It is the margin of variation in concentration of the gases that is alarming. For example, CO_2 concentration in the atmosphere stayed the same from 1800 years at an average of only 280ppm while in the last 200 years it has risen to a maximum average of 380ppm. Also, the variation between the AD1800 years and last 200 years is disproportionately insurmountable. For instance, there was minimal range of only 15ppm in the 1800 period but this changed in the last 200 years to become a range of 100ppm. This produced a net surplus of about 75ppm of atmospheric CO_2 concentration within the last 200s year of the past AD2000 years. The same goes for N_2O whose range was also 15ppb within the 1800 years dispensation, but later suddenly shot up to a range of 50ppm, producing a net surplus of 35ppm of N_2O concentration in the atmosphere during the last 200 years. Methane equally became more abundant in the atmosphere. From the range of 250ppb during the AD1800 years to a phenomenal 1,150ppb range variation in the last 200 years, producing a net surplus of about 900ppbs of CH_4 concentration in the atmosphere. This insurmountable increase in concentration of the three gases in the atmosphere in the past 200 years in comparison to the 1800 years before clearly indicates an abnormality that seems to have transpired after the late 1800s. It is also interesting to note that this period which marks the sharp rise in CFC gas concentration coincides with the industrial revolution. The industrial revolution in Europe began in the late 1700s and by the late 1800s, various factories had spread across most European countries and CFC gas emission started skyrocketing. As industry and urbanization gained momentum in human society, so did CFC emission. Therefore, there is a correlation between human development and climate change embedded in the mutual relationship between industry and increased emission of CFCs into the atmosphere.

Apart from that, the change in land-use from simplistic Neolithic (agrarian) rural communities to the complex industrial urban communities, has seen a general depletion of the two most significant components of the eco-system -land and water. Research has shown that within the past 200 years, land and water resources have drastically reduced. This has compounded the problem of CFC emission thus expediting climate change. The figure 13 shows a summary diagram of how factors combine to produce an atmosphere of increased CFC concentration. According to the findings, the use of fossil fuels to run industry has significantly changed the use of land and water. This means that fossil fuel consumption due to industry has reduced land

and ocean sinks. This in turn has resulted in proliferated rise in CFC concentration in the atmosphere predisposing the earth to global warming and climate change.

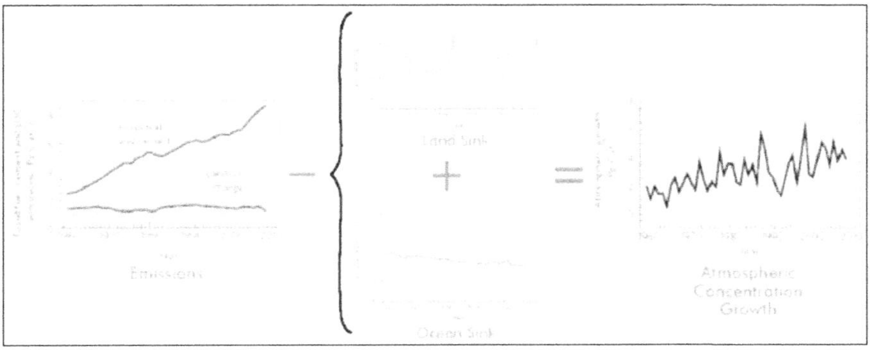

Figure 13. Showing the interplay of Industry and land-use change leading to CFC gas atmospheric concentration

Source: National Research Council, 2011a

2.5.4 Man's Disturbance of world Ecology and contribution to Climate Change

Unlike in Neolithic times, the advent of industry, has seem mankind grossly disturb several aspects of ecology and climate. Among the critical elements that mankind has affected in ecological balance in vegetation cover or more specific -forest cover. The study sought to uncover the extent to which human activity has removed or replenished forest cover on earth. In order to establish this, a 67-year period was selected from the year 1948 when more concrete data on global forest cover was recorded by FAO to the year 2015 when the last consolidated forest assessment report was released.

2.5.4.1 A Timeline Ratio of Natural Vegetative Spatial cover against Brown Spaces

In order to comprehend the gravity of the ecological transformation across the globe, it is imperative to consider the ratio of forest cover relative to how much cover is removed within a certain period. Such an analysis provides insight into the trends of how deforestation has reduced forest cover in the progress of time. Finding of the study are shown in the table 4 indicating the trend of global forest cover and how it has been depleting over time.

Table 4. Global Ecological Deficit in a Timeline Ratio of Vegetative Cover against its removal

Year	*Global Forest Cover (Billion Hectares)*	*Net Global Forest Change (Billion Hectares)*	*Annual Net Change*	*Global Brown Spaces*	*Ecological Deficit*
1948	4.0	-	Not Reported	Unclear	Unclear
1953	3.8	**-0.2**	Not Reported	Increased	Yes
1958	4.4	**+1.4**	Not Reported	Decreased	No
1963	4.1	**-0.3**	Not Reported	Increased	Yes
1988	3.6	**-0.5**	Not Reported	Increased	Yes
1990	3.4	**-9.9**	Not Reported	Increased	Yes
1995	3.4	**-11.3**	Not Reported	Increased	Yes
2000	3.8	**-9.4**	-0.18	Increased	Yes
2005	3.8	**0**	-0.11	Increased	Yes
2010	4.0	**+0.2**	-0.08	Decreased	No
2015	3.99	**-0.01**	-0.08	Increased	Yes

Source: FAO Assessment Reports

For the most part, the period from the year 1948 to 2015 has seen a sustained erosion of vegetation across the globe. The direct inference of this is generally ecological deficit. Approximately, out of the 10 dispensations observed, 80% of them showed ecological deficit and only 20% showed surplus. This significant (80%) difference in ecological deficit indicates that, the world is consuming more of vegetation than it is replenishing. A typical example is about countries on the African continent which are now exploiting forests at an increasing rate. According to the FAR (2005), in Africa, overall progress towards sustainable forest management in the continent appears to have been limited during the last fifteen years. There are some indications that net loss of forest area has slowed down and that forest areas designated for conservation of biological diversity has increased slightly. However, the continued, rapid loss of forest area – the largest of any region in the world during this 15-year period – is particularly disconcerting. This is a similar story for even other parts of the global south such as India, South America and some parts of the Middle East.

2.5.4.2 Net Loss of Global Forest Cover over a 67-year period (1948 to 2015)

There has been a net loss of forest cover between the year 1948 to 2015. This net loss is as a result of the rate at which forest cover has been eroded and seemingly replaced worldwide. The trends reveal a pattern of fluctuation in removal and replacement of forest cover. However, according to the findings of the study from an assessment of past records, the final correlation yields a negative t-Statistic value of -0.185 which indicates an overall net loss. The table 5 shows the findings after running a One-tailed, paired Two-Sample Means t-Test which yielded a significant result. Findings were that the p-value was 0.429 which was less that the Pearson correlation of 0.458. Also, the t Statistic (-0.185) is less than the t Critical (1.860) at one tail. Therefore, from the foregoing, the result shows a negative correlation indicating a global net loss of forest cover in the period between 1948 and 2015.

Table 5. Showing a Paired Two Sample Means t-Test of forest cover from 1948-2015

t-Test: Paired Two Sample for Means	Forest Cover Billion Hectares	Forest Cover Billion Hectares
Period of change (years)	*before*	*after*
1948 - 1953	4	3.8
1953 - 1958	3.8	4.4
1958 - 1963	4.4	4.1
1963 - 1988	4.1	3.6
1988 - 1990	3.6	3.4
1990 - 1995	3.4	3.4
1995 - 2000	3.4	3.8
2000 - 2005	3.8	3.8
2005 - 2010	3.8	4
2010 - 2015	4	3.99
Mean	3.81111111	3.83222222
Variance	0.10611111	0.10959444
Observations	9	9
Pearson Correlation	*0.4576021*	
Hypothesized Mean Difference	*0*	
df	*8*	

t Stat	*-0.1851479*
P(T<=t) one-tail	***0.42886044***
t Critical one-tail	*1.85954804*

Source: Generated by Author, 2022

Therefore, from the foregoing, the Null Hypothesis (H_o) which states that there is no net loss of vegetation over built environment in the world between the year 1948 to 2015 is rejected, while the Alternative Hypothesis (H_1) which states that there is a global net loss on forest cover over built environment between the year 1948 to 2015 is accepted. Consequence, such a significant net loss ought to be checked, lest the trend continues and the global loss of vegetation cover increases exponentially with the passage of time. And if this happens, the ramifications of ecological imbalance will ravage the planet at dangerous levels.

2.5.5 Global Ramifications of Ecological Disturbance and Climate Change

2.5.5.1 The Ecological ramifications of Climate Change

The ecological implications of climate change can be seen in the drastic change of average temperature. Findings reveal that temperatures in the stratosphere have been dropping, while temperatures in the troposphere have been rising above the normal average. The figure 14 shows the cooling trends in the stratosphere and the warming trends in the troposphere in the period between 1960 to 2010. This means that those temperatures in the upper part of the atmosphere are significantly dropping below the normal. Meanwhile, temperatures in the lower part of the atmosphere with which we mostly interact are rising to disproportionate levels. The rise of temperature in the troposphere has led to the predicament called global warming. Generally, all eco-systems on earth directly interact with the troposphere. Infact, the earth-atmosphere system basically comprises the land, water, vegetation and all the weather elements of the troposphere. Therefore, a rise in temperature affects all the aspects of ecology.

An analysis of the collective impact of different factors in the earth-atmosphere system shows a net effect of human activity as being the largest contributor to global warming. The study established that there are some factors that influence of cooling and warming of the atmosphere. Among these factors include the emission of greenhouse gases, depletion of ozone, availability of water vapor, surface reflexivity, particle and cloud reflexivity. According to the findings, greenhouse gases emitted largely through human activity (industry and urbanization) is the greatest contributor to global warming. The gases which mainly include CO_2, N_2O, and CH_4 have high HRC, and therefore trap heat in the atmosphere. These gases can stay within the atmosphere for a span of between 10 to 30 years and continue to absorb and contain heat for as long as they are present.

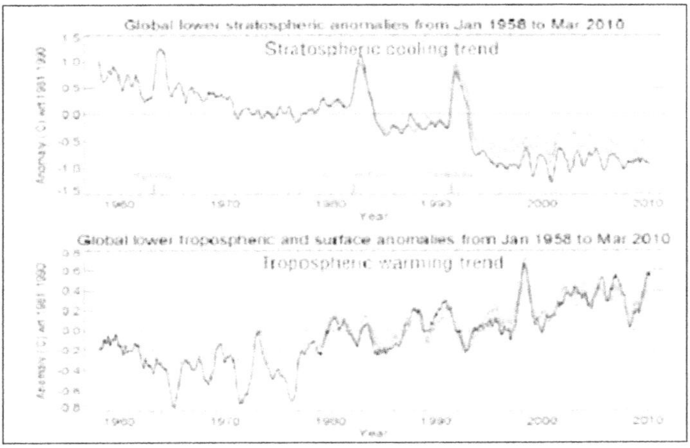

Figure 14. Showing the global cooling and warming trends in the atmosphere

Source: National Research Council, 2011a

Apart from the three CFCs aforementioned, other gases such as Carbon Monoxide (CO) and Sulphur Dioxide (SO_2) which are continuously released from mines and manufacturing industries are largely responsible for the depletion of ozone (O_3). These gases tend to split O_3 by extracting an oxide (O) from it so that it becomes oxygen (O_2). The findings show that depletion of O_3 is creating increased warming conditions due to increased penetration of Ultra-Violet Radiation (UVR) and other thermal energy waves into the atmosphere. Water vapor content has also reduced as temperatures rise further. Surface reflexivity has been affected by two factors with counteracting effects. For instance, land-use that is now mainly characterized by urban built environment has greatly increases reflexivity, thereby raising temperatures. Additionally, soot from carbon which collects on snowflakes reduces reflexivity and generally create warmer conditions in the atmosphere. The figure 15 shows a summary of the interplay of natural and human influence upon the balance of the earth-atmosphere system. The figure shows how human beings are altering the natural course of atmospheric temperature equilibrium. Given a collection of all the factors at play in the earth-atmosphere system, there interaction has ultimately led to an energy net gain from both natural influences and human activity. Most importantly human activity has transcended its impact on global warming against natural factors on a scale of 1: 2.5. This significant difference in the ratio to which human activity is far exceeding natural influence in global warming places the catastrophe of climate change solely upon human beings. The greatest contributor to the earth-atmospheric ecological imbalance is greenhouse gas emission and these are a direct by-product of industry and urbanization.

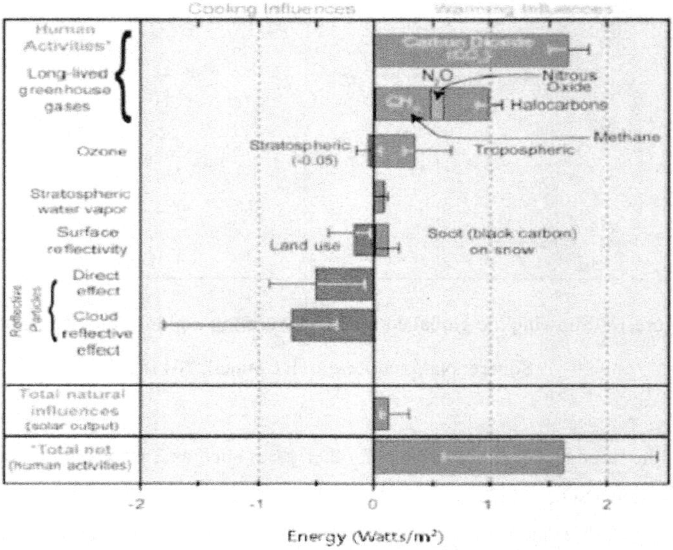

Figure 15. Showing the interaction of natural and human influence on the average atmospheric temperature

Source: National Research Council: 2010d

Apart from the atmospheric part of the earth-atmosphere system, climate change has also adversely affected the earth part of the system. In this way, the land, water and vegetation resources have been altered in composition and functionality. For instance, sources indicate that oceanic potential Hydrogen (pH) has been dropping insurmountably. According to the studies, between the years 1989 to 2007 the oceanic pH within the depth of about 0-30 meters have dropped from an average of 8.12 down to about 8.08. This poses a danger not only to aquatic but terrestrial life as well. Lower pH entails decreased alkaline conditions and increased acidity, a predicament with high chances of further repercussions on other parts of the environment and biodiversity. Apart from the decrease in pH, the quantity of seaweed has also been increasing following the rise of CO_2 content in the atmosphere. For instance, studies show that the consumption of CO_2 by seaweed between the year 1989 to 2007 has increased from a volume of 330 units to about 360 units, giving a net excess of 30 units. This scenario if depicted in the figure 16. showing how seaweed is steadily increasing alongside CO_2 and how the oceanic pH is also decreasing.

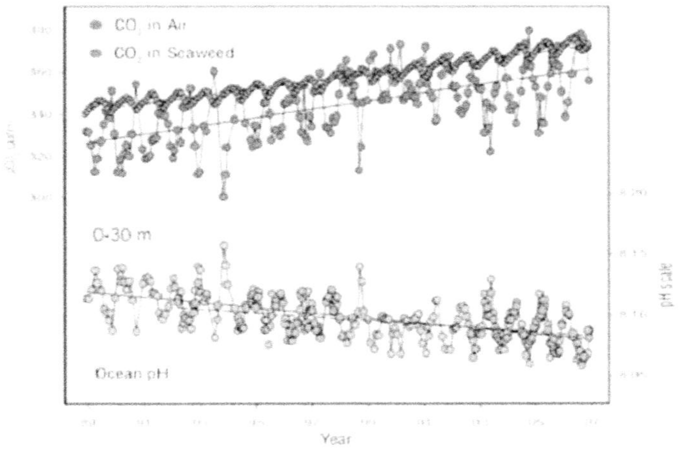

Figure 16. Showing the escalation of seaweed alongside CO_2 and dropping of oceanic pH

Source: National Research Council: 2010d

The increase in seaweed following increased volume of CO_2 means that invasive species are multiplying at an alarming rate threatening natural species. This entails that aquatic life is endangered and so are other aspects of vegetation and animal life that survive on aquatic creatures. This challenge rallies alongside acidity of the water bodies that pose a threat to life in any of its form that survives at an average pH. The general shift in ecological composition of different ecosystems is also fast changing biodiversity. Findings also reveal that, subtle shifts in gaseous composition, pH levels and temperature degrees are resigning the ecological footprint. The predicament is recreating natural habitation into something unfamiliar to the existing biodiversity. The current natural environment being suited to the excising conditions are likely to experience a struggle to adapt to the new changes in ecological construction. This will entail that other species may emerge that are better suited for the new conditions and this leads to the development and evolution of foreign/invasive species. It may not be so clear as to what the foreign species adapted to the new ecology can do but at least some studies have revealed that most of such species are invasive and terribly harmful to the current species. A typical example if the black beetle found the Californian forests which has ravaged the vegetation there to the brink of extinction. The beetle literally feeds on entire branches removing chlorophyl and sucking the life out the plants. The photo in figure 17. shows how the black beetle has ravaged a forest in California -USA.

Figure 17. Showing how the invasive Black Beatle has ravaged a forest in California

Source: National Research Council: 2010d

In Africa, Zambia in particular, number of invasive species have evolved apart for the Kafue weed such as water hyacinth. These weeds have mainly been a source of the destruction of crops in the areas where they have appeared. They have made agriculture almost impossible in the areas where the Kafue River passes. Other invasive species in Zambia include the Mimomsa Pigra plant from South America, which is ravaging wildlife especially in the area around the Kafue flats.

2.5.5.2 Social-economic ramifications of Climate change

Climate change has phenomenal impact upon human life. The unprecedented shifts in weather patterns have produced catastrophe such as floods and outbreak of disease in several parts of the world. Global warming has not only raised the sea levels but created conditions favorable for certain fleas to survive. Various microbes harmful to human health such as the bacteria and viruses like the Chaleo Vibrio (Cholera Virus) has in modern history ravaged Europe, Asia and Africa (Ndebele-Murisa *et al.*, 2020). Although Europe and Asia contained the situation in the early 1900s, Africa and India seemed to have failed for the most part to contain it. The predicament seems to affect the two localities on earth perhaps due to the abundance of informal settlements or slums in urban areas. It is these spaces of urban blight where bacteria and viruses that cause water bone diseases are commonly found. This is probably due to their unsanitary conditions. And so, the challenge of informal settlement becomes top priority in forwarding the urban agenda on several international forums. In the most recent years, during the period between 2020-2022, a strange virus called Corona Virus -19 (COVID) emerged and enveloped most countries with cooler climatic conditions. It was viewed by the WHO that the virus flourishes best in temperate regions where temperatures where relatively quite low. Nonetheless, it is interesting to note that the same virus as well as its hybrid still spread to countries in the tropics (mostly found in Africa) and infected large populations as well (Chigbu and Onyebueke, 2021). This is indicative that temperatures in Africa during the cold season have dropped below the normal. Throughout the world, millions of people lost their lives at a rate never experienced before since the last outbreak of small pox. Perhaps, COVID-19 was the first tip of the iceberg showing us the fate a future whose ecology is completely distorted.

Overall, we can also say that climate change has a cost implication upon the lives of people. the economic implication is created by the loss of human resource, infrastructure breakdown, and the disruption of trade and commerce. The COVID -19 pandemic literally halted the peak performance of business. Many people lost employment during the 2020-2021 lockdown (Chigbu and Onyebueke, 2021). Generally, the laws of supply and demand were grossly disturbed and this was typically seen in the agriculture sector in United States of America, where gallons and gallons of milk was disposed-off into canals as refuse due to lack of demand. The world was clearly headed for an economic meltdown similar to that of 2008.

2.6 THE FUTURE OF GLOBAL ECOLOGY AND CLIMATE CHANGE

Findings show the pillars upon which human survival and advancement is premised upon in order to project like-hood of ecological depletion, thus, unveil the future of climate change. Findings also reveal that humanity is grappling with a dichotomy of population change and poverty across the globe. These are matters that have grave ramifications upon ecological preservation (UN-Habitat, 2022; Giddings, Hopwood and O'Brien, 2002). The two matters are also significantly part of the motivation behind human technological advancement and the quest to develop. As a matter of fact, the two are among the Sustainable Development Goals (SDGs). Findings also show that, in an attempt to attain such goals, the global community is contending with a complex situation whereby, development naturally demands supply from the ecology, while the ecology also demands replenishment. Failure to reconcile the two is likely to breakdown ecological stamina and unleash all the dangers of climate change. Therefore, the concept of sustainable development becomes handy but also complicated at the same time. In order to understand the potential of population change in the matters of climate change, the section that follows maps out populations change across the world. This is to give light into the situation obtaining in the world about how much needs to be addressed to aid the status quo for most parts of the world.

2.6.1 Population Change in the trajectory of world Ecology and Climate Change

The world map in figure 18 shows the population change across the world in the period between 1990-2013.

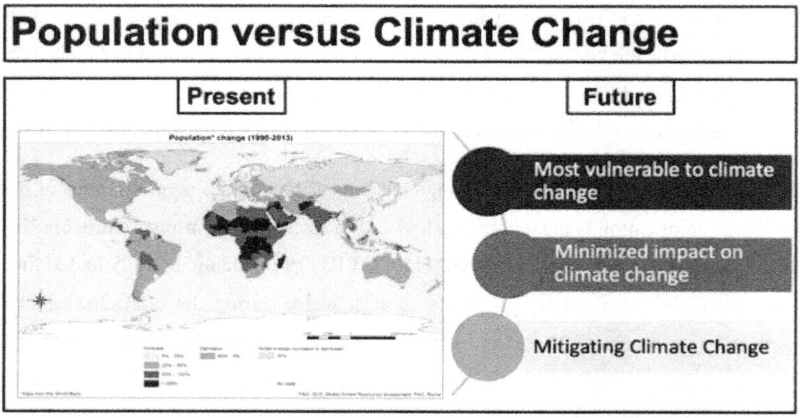

Figure 18. Showing global mapping of population change against potential climate change

Source: Adapted from FAO Assessment Report, 2015

According to map evidence, much of the world falls in the category of between 25 to >100 percent increase in population and only less than 25 percent have minimal, zero, or negative growth. This shows the recent historical trajectory of world population dynamics. Populations are generally expanding and this is increasing the pressure on development agendum for many nations. Extremely few (<5 percent) countries which are located in the Scandinavian world are the ones that have attained an impressive global trajectory with a promising future for decreased pressure upon natural ecology. If the majority are still striving with efforts to reach zero or negative increase in population, then the challenge of population pressure upon the ecological footprint continues. This in turn culminates in ripple concerns over the climate change issues. All the parts of the world with increase in population are mapping out areas of potential ecological degradation. And unless measures are put in place, the world still risks chances of losing the ecological footprint.

2.6.2 Poverty in the trajectory of world Ecology and Climate Change

As mentioned in an earlier section, the other factor motivating ecological erosion is the relentless quest to alleviate poverty across the world. Poverty is a social-economic ill in human society and it is one of the main issues being addressed by global communities. Findings show that, this zest to make as many people access basic social and economic needs has a telling impact on the environment because the obvious instruments are industry and technology. However, these engines of mass production escalate the consumption of natural resources and serve to accelerate the rate of ecological degradation and climate change. As at 2015, the prevalence of poverty in the world according to FAO is shown in the figure 19. As shown in the findings, literally the entire global south including parts of Central Africa, Central America, India and the islands of East Indies, except for Australia is still ravaged by poverty and this state of affairs calls for more implementation of developmental plans. If there are still traces of large portions of land ravaged by poverty, then there is still greater potential for developers to deplete the ecology.

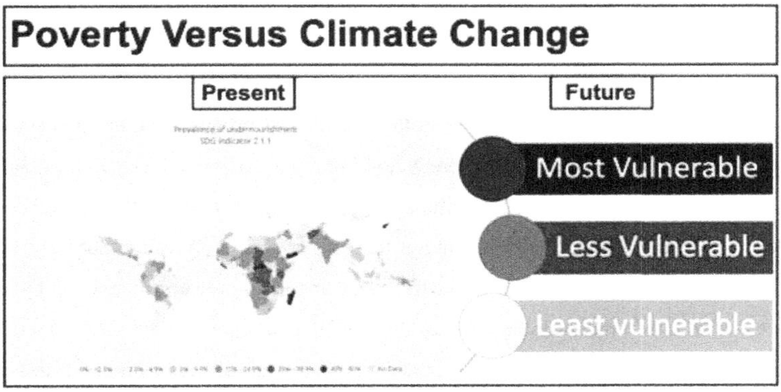

Figure 19. Showing mapping of world poverty against potential climate change

Source: Adapted from FAO Assessment Report, 2015

https://nap.nationalacademies.org/read/14673/chapter/2#3

It is the exploitation of natural resource which further robs it of the natural elements needed to support healthy conditions on the planet. And if this potential for further exploitation exists, then the future of climate change is still in limbo. Today, there are not so many ways of alleviating poverty except through developmental agendas that resort to depletion of natural resource. Strides are being initiated to develop ways of combating poverty with minimal damage to the ecology. Nonetheless, such strides are still insufficient to reach the mass number of people who still need to be salvaged from the shackles of poverty especially in Africa.

2.6.3 Economic development in the trajectory of world Ecology and Climate Change

Findings of the study have established that there is a correlation between human development and climate change. Human development has dictated the pace of land use change and led to increased emission of greenhouse gases. Primarily change of land-use in the past 200 years has been characterized by the establishment of industries and the spread of urbanization. Compounding this land-use changes is the continued rapid growth of human population cross most parts of the world. The meaning of such population growth in the dynamics of climate change is that, industry will continue releasing of greenhouse gases and temperatures are likely to continue rising (Harvey, 2018). The double flanked problem is that, industry and urbanization consume massive amounts of natural resources in terms of land, water and forest, while at the same time emitting greenhouse gases. This means that ecological erosion and pollution are happening at phenomenal rates around the globe at phenomenal (Ibid, 2018).

Another critical component of the ecology has been introduced in the mix of land and oceanic degradation -forest cover. This is yet another aspect of the ecology that has been significantly impacted by human development. Again, both industry and urbanization has grossly reduced forest cover across the world. Today, there is nothing exploited by industry more than forests. Many sectors of the economy demand timber, and so, the lumbering industry has received more attention that several others. The challenge in the industry is that trees are not only exploited to produce manufactured goods in industries but are also indiscriminately cut down for clearance to establish industrial plants or even settlement as well as shifting cultivation (Cudworth, 2005). This makes forest cover one of the most threatened aspects of the ecology. The global trend across history shows that brown spaces are not only draining marshes but swallowing up forest cover.

The world map today shows how the built environment is encroaching green spaces and overtaking vegetative cover. Although urban planning seems to be taking sustainable planning concepts to include green spaces, this kind of integrated planning has largely been ignored in the past 200 years by many communities in the world (UN-Habitat, 2020; Mori and Christodoulou, 2012). This is especially common in the global south were unplanned or informal settlements are rife. It is an international urban challenge which the UN seeks to address in this 21st century but so far has recorded minimal success (UN-Habitat, 2022).

Despite efforts to transform planning into a means of establishing smart cities to incorporate greening, brown development continues to override in many cases. For example, shopping malls in urban centers have for some time removed vegetation and paved vast spaces depriving land sinks and vegetative cover. There are now new approaches and concepts in planning that seek to address such issues. The figure 20 shows how brown spaces are predominant the green spaces.

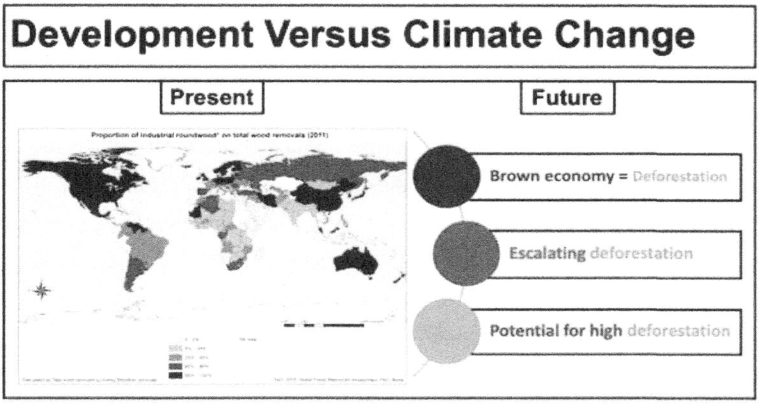

Figure 20. Showing spaces of brown development across the globe against their potential for deforestation

Source: FAO Assessment Report, 2015

All the spaces mapped out with brown are areas of high deforestation due to peak industrialization. Those are spaces most vulnerable to desertification. Nevertheless, they are also communities that have taken matters of forest management more seriously that the spaces mapped-out in brighter colours. It also seems that the communities that are less developed have more forest cover compared to that are developed. Perhaps the developed countries depleted their forests in a bid to attain development. And this becomes the trap thesis of poverty. If development will come to the global south and poverty should be eradicated, then deforestation must rise to a certain threshold. And yet with deforestation also comes climate change and other ecological ramifications.

2.7 CONCLUSIONS

The advancement of human so-called civilization from Prehistory, Neolithic, Iron, Technology, Industry, and Artificial Intelligence are forerunners of quantum ecological transformation.

Prehistoric and Neolithic times have shown a more natural/normal curve of ecological cycles. The iron age increased productivity and structured human settlement of human without significantly altering ecology. Agriculture consumed forest and iron-working expedited this but also without significantly disturbing ecology. Arguably, the industrial revolution has contributed the greatest deal to ecological disturbance and climate change. The late 19th century began to realize higher levels of CFCs and temperature in the atmosphere. This period upset a chain reaction and other parts of the ecosystem such as water and land reacted to the atmospheric restructuring. All these changes inevitably began to impact biodiversity in terms of flora and fauna both in terrestrial and aquatic domains. The industrial revolution has spread across the globe and its concomitant ecological ramifications have followed it. The industrial revolution has also precipitated the birth of further technological advancement called artificial intelligence. This is the current trend of human achievement and is dictating the pace of productivity and economic development. Nonetheless, economies of various countries that are still developing must be reoriented on the path of sustainability to stop the possibility of future catastrophe. Therefore, urgent action to curb the escalating ecological disturbance through integrated conceptual planning and initiatives from the SDGs ought to be quickly implemented to protect the planet from impending doom.

ACKNOWLEDGEMENTS

I thank the Almighty God for the opportunity to write this paper and to make my submission to the CiVEJ and AIRCC. Firstly, I thank the Academia forum for the notification online which alerted me about the chance to make the submission. Secondly, I also thank Mrs. Sinkamba J.N., Dr. Muchanga M., Miss Chilufya K., Miss Chirwa F., Mr. Kaulule C., Mr. Phiri G. and Miss Tembo G., for their continued support in my scholarly and academic endeavors. I am also grateful to Dr. Chibamba D., Eld. Mwika G., and Mr. Chirwa R., for their encouragement as I wrote. I thank my Eco-club team for their zeal for community development which continues to be a beckon of hope rekindling the drive in me to help informal settlements. I lastly thank my family for moral support.

REFERENCES

[1] Arlt, H. and Daviau, D.G. eds., 2009. *Culture, Civilization and Human Society–Volume II* (Vol. 4). EOLSS Publications.

[2] Bartke, S., Forster, T., Githiri, G., Jering, A., Kago, J., Schlimmer, S. and Sietchiping, R., 2021. The UN-Habitat Urban-Rural Linkages Guiding Principles: Assessment of the Adoptability to Topical Land Management Challenges in Germany, Kenya and Tanzania. In *International Yearbook of Soil Law and Policy 2019* (pp. 369-398). Springer, Cham.

[3] Bostley. A., 2019. *Basics of Research Design: A Guide to selecting appropriate research design*. 6. 76 – 89.

[4] Chibuye, 2014. Interogating Utban Poverty Lines. Oxford

[5] Chigbu, U.E. and Onyebueke, V.U., 2021. The COVID-19 pandemic in informal settlementsre) considering urban planning interventions. *The Town Planning Review*, *92*(1), pp.115-121.

[6] Clark, G., 2014. The industrial revolution. In *Handbook of economic growth 2*, 217 262. Elsevier.

[7] Dovey, K., Shafique, T., van Oostrum, M. and Chatterjee, I., 2021. Informal settlement is not a euphemism for 'slum': what's at stake beyond the language? *International Development Planning Review*, *43*(2).

[8] Evidence, Impacts, and Choices: PDF Booklet. 2015. Washington, DC: The National Academies Press.

https://doi.org/10.17226/14673.

[9] Food and Agriculture Organization of the United Nations. 2020. *Global Forest Resources Assessment.*

https://www.fao.org/forest-resources-assessment/remote-sensing/global-ecological-zones-gez-mapping/en/

[10] Giddings, B., Hopwood, B. and O'brien, G., 2002. Environment, economy and society: fitting them together into sustainable development. *Sustainable development*, *10*(4), pp.187-196.

[11] Habitat, U.N., 2019. *The strategic plan 2020-2023.*

[12] Habitat, U.N., 2022. *World Cities Report 2022: Envisaging the Future of Cities.*

[13] Habitat, U.N., 2018. *Tracking progress towards inclusive, safe, resilient and sustainable cities and human settlements*

[14] Hale, A.J., Ricotta, D.N., Freed, J., Smith, C.C. and Huang, G.C., 2019. Adapting Maslow's hierarchy of needs as a framework for resident wellness. *Teaching and learning in medicine*, *31*(1), pp.109-118.

[15] Harvey, L.D., 2018. *Global warming.* Routledge.

[16] Held, I.M. and Soden, B.J., 2006. Robust responses of the hydrological cycle to global warming. *Journal of climate*, *19*(21), pp.5686-5699.

https://www.scribbr.com/statistics/t-test/

[17] Jorgenson, A.K., 2003. Consumption and environmental degradation: A cross-national analysis of the ecological footprint. *Social Problems*, *50*(3), pp.374-394.

[18] Klasen, S., 2018. Human development indices and indicators: A critical evaluation. *Human Development Report Office Background Paper*, *1*

[19] Labat, D., Goddéris, Y., Probst, J.L. and Guyot, J.L., 2004. Evidence for global runoff increase related to climate warming. *Advances in water resources*, *27*(6), pp.631-642.

[20] Muhanna, E., 2017. The World in a Book. In *The World in a Book.* Princeton University Press.

[21] *National Research Council: 2011d*

[22] *National Research Council: 2010d*

[23] Nchito, W.S., 2007. Flood risk in unplanned settlements in Lusaka. *Environment and Urbanization*, *19*(2), pp.539-551.

[24] Ndebele-Murisa, M.R., Mubaya, C.P., Pretorius, L., Mamombe, R., Iipinge, K., Nchito, W., Mfune, J.K., Siame, G. and Mwalukanga, B., 2020. City to city learning and knowledge exchange for climate resilience in southern Africa. *PloS one*, *15*(1).

[25] Pieterse, J.N., 2010. *Development theory*. Sage.

[26] Rashid, A., Irum, A., Malik, I.A., Ashraf, A., Rongqiong, L., Liu, G., Ullah, H., Ali, M.U. and Yousaf, B., 2018. Ecological footprint of Rawalpindi; Pakistan's first footprint analysis from urbanization perspective. *Journal of Cleaner Production*, *170*, pp.362-368.

[27] Rees, W., 2018. Ecological footprint. In *Companion to environmental studies* (pp. 43 48). Routledge.

[28] Rymarczyk, J., 2020. Technologies, opportunities and challenges of the industrial revolution 4.0: theoretical considerations. *Entrepreneurial business and economics review*, *8*(1), pp.185-198.

[29] Scarre, C. and Fagan, B.M., 2016. *Ancient civilizations*. Routledge.

[30] Singer, S.F., 2006. *Unstoppable global warming: Every 1,500 years*. Rowman & Littlefield

[31] Vlek, C. and Steg, L., 2007. Human Behaviour and Environmental Sustainability: Problems, Driving Forces, and Research Topics. *Journal of social issues*, *63*(1), pp.1-19.

[32] Wengrow, D., 2018. *What makes civilization? The ancient near East and the future of the West*. Oxford University Press.

[33] Xu, M., David, J.M. and Kim, S.H., 2018. The fourth industrial revolution: Opportunities and challenges. *International journal of financial research*, *9*(2), pp.90-95.

[34] Yumashev, A., Ślusarczyk, B., Kondrashev, S. and Mikhaylov, A., 2020. Global indicators of sustainable development: Evaluation of the influence of the human development index on consumption and quality of energy. *Energies*, *13*(11), p.2768.

CHAPTER THREE: ENHANCING RESILIENT HOUSING IN INFORMAL SETTLEMENTS OF LUSAKA

Abstract

Resilient housing in informal settlements implies adaptability for a sustained access to proper housing. Establishing the ability of housing to withstand vulnerabilities in informal settlements can help the majority of people in the world because most people live in urban informal settlements, which are, places of unplanned houses below minimum standards as defined by the Hedonic House Pricing (HHP). This study considers Lusaka's Kalingalinga and Mtendere East informal settlements where, a total stratified sample of 60 households, 30 from each area were studied. Findings reveal that; intention to extend houses, improvision, social cash transfers, and involvement in community action are the most important grassroot initiatives in the two settlements. A one-way ANOVA 2 sample variant test at alpha level 0.5, 1 tailed, indicates greater potential for enhancing resilient housing in the two settlements using variables such as; intention to extend, space for upward adjustment, proximity to service pipes, and improvisions. Meanwhile, despite such potential, the statistical (ANOVA) test has also revealed that the existing housing situation (that is below minimum planning standards) has not improved despite the existence of such grassroot initiatives. Therefore, the paper has sought to resolve this aberration by reconciling grassroot initiatives with institutional arrangements using an integrated planning approach known as Polycentric Planning Strategy (PPS). To achieve this, a system of interlinkage has been mapped out, connecting existing grassroot initiatives and structures with planning authorities and other stakeholders to raise solutions for enhanced resilient housing in informal settlements.

Keywords: *Resilience, Grassroot Initiatives, Housing, Integration*

3.1 INTRODUCTION

Owing to unprecedented shifts of global climate and economy which affects housing for majority population in recent times, this paper endeavors to highlight the housing challenges existent in Kalingalinga and Mtendere East settlements of Lusaka city in Zambia due to the lack of urban resilience. It labors to show case the resilient housing initiatives at grassroot level among the residents, then measures the degree of potential that the two areas sustain for enhanced resilient housing. After that, the paper assesses the extent to which the grassroot initiatives are adapted to the housing challenges. Through this, it seeks to unveil lacunars in the grassroots initiatives to indicate where deliberate panning can play a role to enhance resilience. Furthermore, it establishes how spatial planning can enter upon the shortfalls of grassroot initiative to boost them and make them more effective in fostering resilient housing. In order to map out such a strategy, the study has adopted the PPS that devolves planning to lower levels of the community. It is worth indicating that, although the concept of resilience may not be well understood by some people at the grassroot level, it is nonetheless something they are already attempting to achieve through some means of adapting their houses to physical, social, and economic shocks within their environment. In their own way, they have formulated what can be referred to as 'grassroot make-shift solutions' to solve their eminent challenges. This paper explores them and emasculates the hidden lacunars and potentialities for enhanced resilience. Most importantly, the paper offers a solution through an integrated planning approach tailor-made for the situation at hand.

Kalingalinga and Mtendere East are informal settlements in Lusaka city characterized by high population density following few houses, most of which are below minimum planning standards. The housing units of these areas reflect on a wider scale the desperate situation preeminent in many informal settlements of the city. Three main problems arise from this state of affairs. One is the limited number of houses accommodating a disproportionate mass of population, and two is the depraved conditions in which the houses are because most of them lack minimum standards. The first challenge culminates into overcrowding in houses which are often relatively small in size. This compromises the people's standards of living in terms of access to sufficient shelter, space, privacy and safety, thereby posing health and security concerns. The second challenge is the dwellers' limited access to adequate water and descent sanitation, which poses health challenges related to water-borne diseases. Thirdly, the limited strong building materials, lack of drainage infrastructure, and limited roofing poses threats of possible calamity in events of natural disasters such as floods, lightning, storms or strong wind. On city-wide scale, the depraved condition of housing in the two settlements has compromised the beauty of the city as it adds to urban blight. The state of affairs in the two settlements shows that people are not living in a space that is safe and habitable.

The aim of the study was to assess the grassroot initiatives that sustain resilient housing in Kalingalinga and Mtendere East so as to establish how they can be enhanced by spatial planning.

The objectives of the study were:

i. to explore the grassroot initiatives of resilient housing in Kalingalinga and Mtendere East;
ii. to assess the extent of adaptability which the grassroot initiatives sustain for securing resilient housing in the two settlements; and
iii. to establish how spatial planning can improve the grassroot initiatives for the enhancement of more resilient housing in the two settlements.

From the foregoing, the hypotheses were:

i. H_0: There is low potential for enhancement of resilient housing in Kalingalinga and Mtendere East settlements.

   H_1: There is high potential for enhancement of resilient housing in Kalingalinga and Mtendere East.

ii. H_0: Existing grassroot initiatives of resilient housing in Kalingalinga and Mtendere East are not well adapted to meet minimum standards of housing.

   H_1: Existing grassroot initiatives of resilient housing in Kalingalinga and Mtendere are well adapted to meet minimum standards of housing.

Kalingalinga is located about seven kilometers east of Lusaka's Town Centre along Alick Nkhata road. It is bordered by Kabulonga in the south, Helen Kaunda in the east, Munali in the north, and Mass Media in the west. Meanwhile, Mtendere East is situated about 10 Kilometers east of the Town Centre, about one and a half kilometers on the south of the Great East Road. It is bordered by NRDC in the north, Ibex Hill in the south, Salama Park in the east and Kalikiliki in the west. The location of the two settlements is as shown on the map in Figure 21.

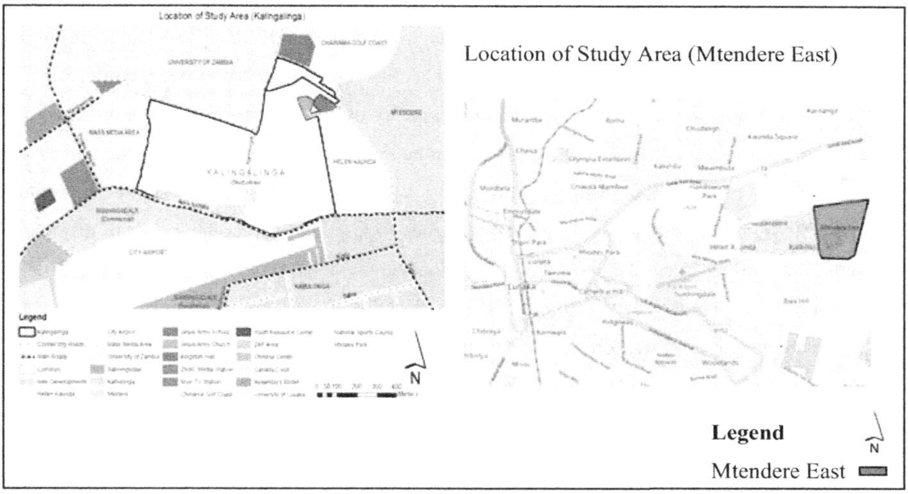

Figure 21. Showing the Location of study area (Kalingalinga and Mtendere East)

Source: Adapted by Author, 2022

Kalingalinga and Mtendere East were chosen as the area of study because they are informal settlements where grassroot resilient housing initiatives exist. The two settlements seem to have potential for development of more resilient housing systems/structures. Most residents in engage in activities of trying to make housing as resilient as possible. Among the commonest means established as a way of coping with the urban morphological reshaping and social-economic stress that has come with urban gentrification in Kalingalinga, is the selling off of properties to relocate from the area (Chikuta, 2016). For Mtendere East, similar things are happening as in Kalingalinga but what prevails the most in the restructuring or extension of the houses.

The first section of the paper describes the concept of housing resilience and contextualizes it to Africa and Lusaka in particular. The next shows the methodology and results of the study before discussing the dynamics of the findings pertaining to how resilient the grassroot initiatives are and how planning can play a key role in establishing that. The paper ends by conclusions and recommendations.

3.2 THE CONCEPT OF RESILIENCE

Urban resilience is the ability of cities/towns to manage and adapt to change through robustness, flexibility, mitigation and adjustment at all levels (Jones, 2017). Meanwhile, Revell (2010) also defines resilience as, the adaptive capacity of a system to respond to unexpected shocks and unforeseen changes. Therefore, it can be said that, urban resilience is the ability of cities/towns to manage and adapt to vulnerabilities due to change. In this regard Sudmeier-Rieux (2014) advances that, cities sustain a tendency to respond to emergent needs of residents based on prevailing circumstances within their local context. To what Sudmeire-Rieux (2014) elaborates, other scholars such as Jones and Suhartini (2014) add that urban responsiveness to eminent challenges within informal settlements does not only happen within formal institutions and structures but also outside formal and clearly spelled out-legal structures, only that they are not resilient enough to the shocks, a situation which the study seeks to address.

Informality in urban areas often springs from uncontrolled urban growth as explained by theories/models which map out the genesis and organic spread of urban phenomenon. Von Thunen's land rent theory well propounds this. The theory states that as the process of urbanization takes place, land rent/value at the core of the city or town increases as more agglomeration economies build up and magnetize within the principal node (Johannes, 2020). This situation attracts more people to the urban area giving birth to overpopulation. Today, Lusaka city accommodates a sum total of over 2 million people and yet it was initially designed by professor Adshed to accommodate only 200, 000 people. Although JICA has expanded the old boundaries to map-out the 'Greater Lusaka', this expansion seems not to yield its desired objective. It is from this backdrop that informality evolves. Mass of the low-income class occupy unplanned settlements that are below minimum standards of decency. SDG 11 indicators reveal that by 2018, 1 billion people in the world were living in informal settlements (UN Reports, 2021).

3.2.1 Resilient Housing in African Urban Informal Settlements

Resilient informal housing entails, housing that is able to adapt itself to imminent vulnerabilities and unprecedented changes that may cause stress/shocks. Such perturbations can make life difficult for informal settlers. According to Akinola (2016) many African cities/towns are unable to accommodate the pressure of concentrated social-economic activities; as a result, majority of the marginalized dwelling in informal urban space often end up establishing their own water and sanitation. A review of studies conducted in Angola, Ghana, Kenya and Nigeria by Akinola (2016) reveals stark separation between the interest of informal settlers and their formal governance structures. He further says that, this results in environmental poverty and urban blight, sometimes warranting evictions and demolition of properties by the state. The author points out that there is no linkage between formal planning institutions and grassroot initiatives. This seems to be the case for Lusaka's Kalingalinga and Mtendere East settlements in Zambia. Akinola's review explored the how ineffective a centralized planning system is and adopts a number of planning strategies including the PPS to show how formal planning can integrate with grassroot initiatives to improve housing in informal settlements. This study also adopts the PPP and adapts it to the context of Kalingalinga and Mtendere East.

3.2.2 Resilient Housing in Kalingalinga and Mtendere East Settlements of Lusaka

Urban resilience has vividly been seen in the limelight of housing in informal areas (Jones, 2017). Lusaka's Kalingalinga and Mtendere East have evolved informal adaptation initiatives to the demands of descent housing. Kalingalinga is an improvement area with its residents having occupancy licenses. The area saw physical upgrading by JICA, and the American Government who installed water kiosks and hand pumps, and service pipes while the LCC surfaced some roads and set-up drainages there. Nevertheless, ever since the upgrading process, most of its dwellers still live under infrastructure void of certain facilities vital to meeting minimum HHP standards spelled out by Tinsley (1997). As a result, the occupants have created resilient strategies to cope with the prevailing situation. Just like Kalingalinga, Mtendere East was also regularized with its occupants being issued land records by LCC. The area then saw the installation of service pipes by the American Government. Inspite of such physical upgrading, the story is the same as that of Kalingalinga since most create residents struggle make their housing as resilient as possible.

On the despite, various actions to accommodate informality have been taken not only by local authorities but also international organizations. Local governments have instituted regularization of these areas. In implementing such, residents of these settlements are either issued with occupancy licenses or land records. Inasmuch as policy and upgrading efforts have been instituted to assist people in informal settings, there is not much aid given to harness the resilient housing initiatives applied at grassroot level. Over time, what has emerged clear are the upgrading efforts by various international organizations such as Swedish and American Governments, and NGOs such JICA and PHPPZ in establishing physical facilities to supplement housing in these areas.

3.3 CONCEPTUAL FRAMEWORK

The conceptual framework of this paper is constructed from a build-up of multiple contingencies; first touching the theory that anchors the study, then its ripple effects that culminate into city growth, social stratification, informality, lack of formal support, indecent housing, resilient local initiative, reconciliation of informality with formal planning, then possible enhanced resilience in informal housing. This construct summarizes the conceptual map of the paper. A summary of how the interlinkages in informality can be harnessed by spatial planning to create solution to the housing problem in informal settlements is shown in the figure 22.

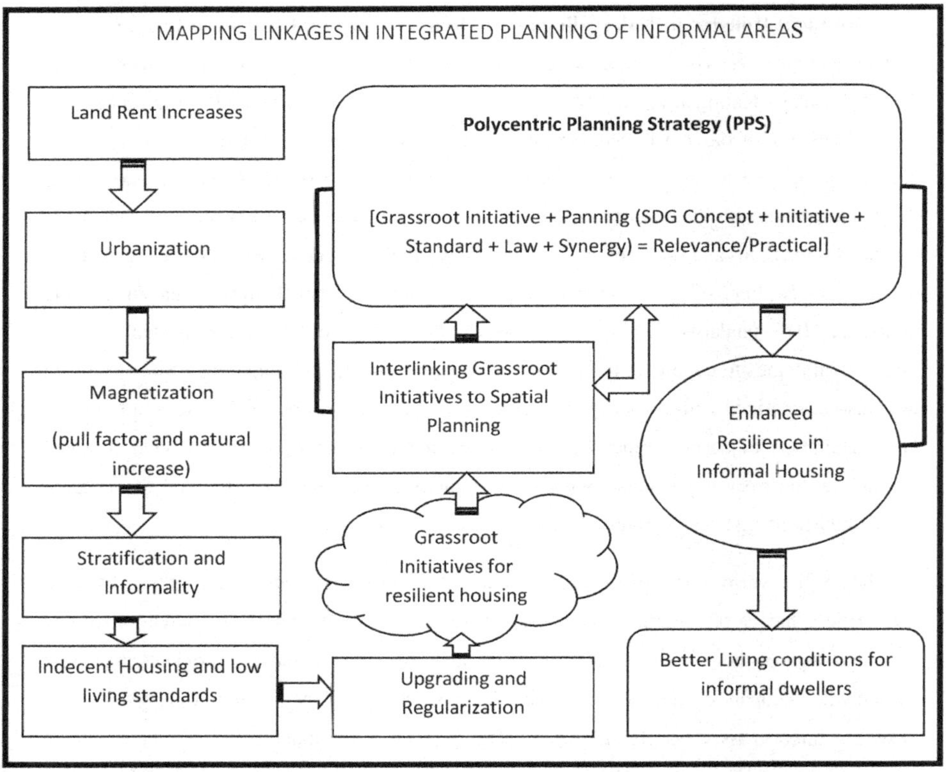

Figure 22. Showing the conceptual framework interlinking grassroot initiatives to spatial planning for enhanced resilience in informal housing

Source: Adapted by Author, 2022

3.4 RESEARCH CONTEXT AND METHODOLOGY

This is a case study design with a mixed (qualitative and quantitative) approach, that considered Kalingalinga and Mtendere East settlements of Lusaka. The design was adopted for the sake of narrowing the focus. This was done to uncover specificities surrounding the grassroot initiatives within the context of an African urban setting. Similar case studies have been done on informal settlements in Africa on Angola, Kenya, Ghana and Nigeria, using the design narrowed to specific contexts. Such studies surfaced sufficient data that was used to formulate linkages between the PPS and grassroot efforts for planning resilience in informal settlements. More to this, the study runs a statistical test to measure the extent of adaptability to vulnerability of grassroot initiatives for sustained resilient housing in the two areas of study.

Random systematic sampling procedure was used on the two settlements, selected because of their nature of informality. Thirty households from each area were selected from main streets. The selection criterion followed one house after every four, for the sake of randomization. The procedure of selecting the housing units was replicated in both areas bringing the total sample size to 60. Five key informants comprised of LCC officials and community leaders and PPPHPZ representatives were also purposively selected. Therefore, the sum total of respondents for the study was 65. Each stratum comprised of six selected households from five selected streets. The households are identified and labelled according to the number in the selection process as h1, h2, h3, ...and so on and so forth, until h30 and replicated in both settlements. The procedure and sample size are shown in figure 23 and table 1.

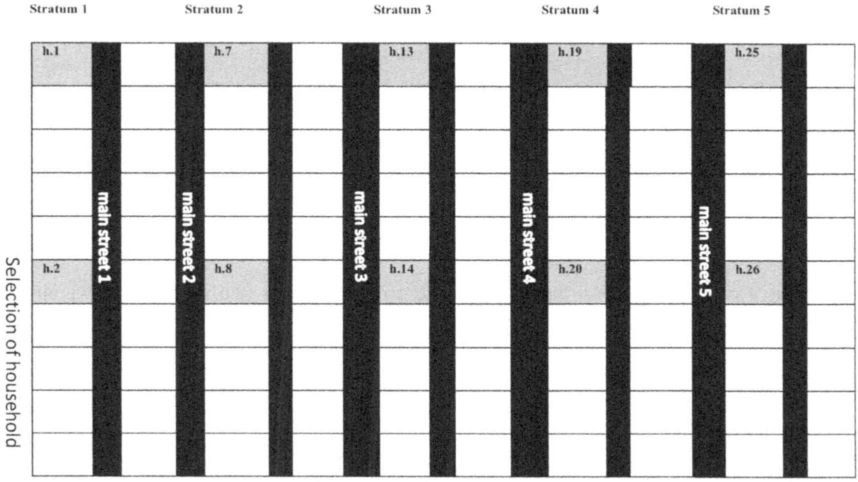

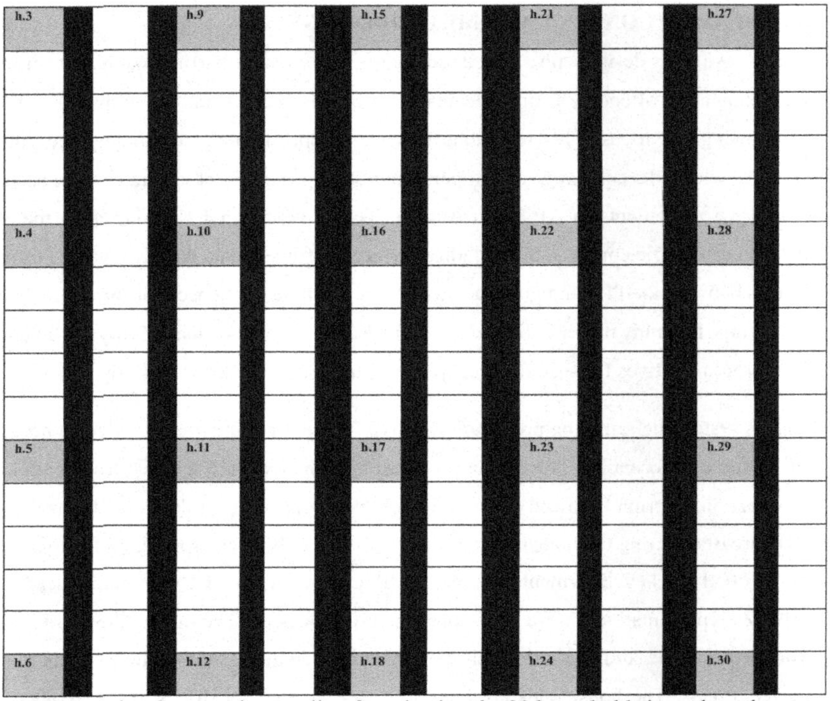

Figure 23. Showing Systematic sampling for selecting the 30 households in each settlement

Source: Adapted by Author, 2022

Table 6. Showing Sample size of key informants and respondents from Kalingalinga and Mtendere East

TARGET	**KEY INFORMANTS**		**HOUSEHOLDS**		**TOTAL**
Sample	**LCC**	**PPHPZ**	**Kalingalinga**	**Mtendere East**	
Sample Size	4	1	30	30	**65**

Source: Author, 2022

The data was collected from the respondents, primarily using two instruments which are a questionnaire and an interview guide patterned after the research questions and hypotheses. Validity of the data was ensured through triangulation of data and the two instruments used enabled cross checking, and for more depth and robustness. Two areas were chosen in order to fill in gaps where other data was missing. There was also physical observation done to ensure triangulation of the data. The respondents were not coaxed or bribed into revealing information. The interviews and photos were taken with their full consent. To assist the launching into the households, consent letters from the institution was used and community leaders were contacted to help introduce the researcher. After the research, the responses will be shredded and burned.

Thematic and content analysis were used on the data. The grassroot initiatives were thematically arranged and the content was explained. The extent of adaptability of the grassroot initiatives to meet minimum housing standards according to the HHP, 'access to social services', CSO (2010) and JCTR (2010) 'measures of household poverty' (Chibuye, 2014) were assessed using a one-way ANOVA 2 variant sample test at alpha level 0.5, 1 tailed. The same test was used to assess whether the minimum housing standards have been reached in the face of grassroot initiatives. Marginal errors were reduced using mathematical constructs such as alpha level 0.5 which is 95% level of confidence. Finally, the integration of grassroot initiatives to formal planning for enhanced resilient housing will be done using the PPS within the context of two settlements.

3.5 RESULTS

This section presents the findings based on the study objectives, namely the existing grassroot initiatives for resilience, extent of resilience due to grassroot initiatives and potential for resilience, as well as the housing situation in the face of existing initiatives.

3.5.1 Respondent Characteristics

3.5.1.1 Gender of Household Heads in the Kalingalinga and Mtendere East

Findings show that there were more (52%) male-headed households than female (48%) headed households in the Kalingalinga and Mtendere East. This is shown in figure 24.

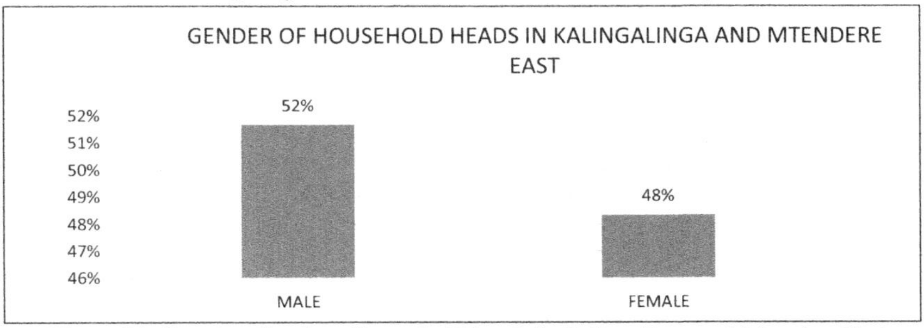

Figure 24. Showing Gender of household heads in Kalingalinga and Mtendere East

Source: Field Data, 2022

3.5.1.2 Household Size and average number of rooms in Kalingalinga and Mtendere East

Most (17%) households constituted of 9 persons per household followed (15%) by 7 and 5 persons per household. This is shown in Figure 25. Meanwhile there are mostly (45%) of 3 roomed houses in the two settlements occupied by an average of 9 persons, meaning that an average of 3 persons occupy a room per house in the two settlements as shown in figure 26.

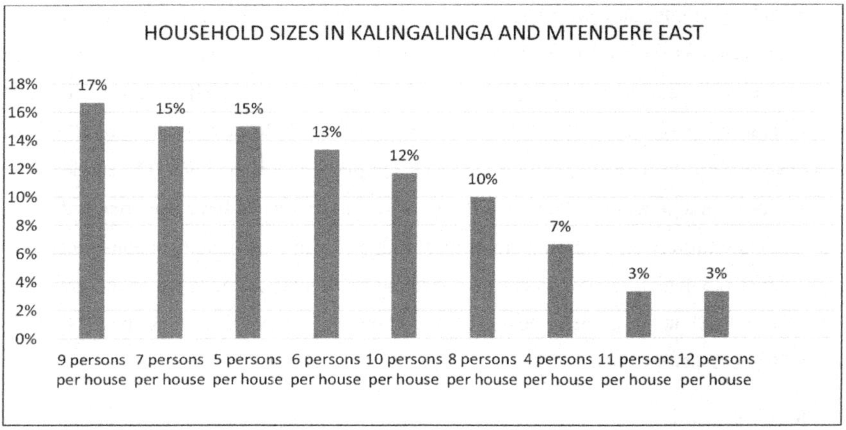

Figure 25. Showing Household size in Kalingalinga and Mtendere East

Source: Field Data, 2022

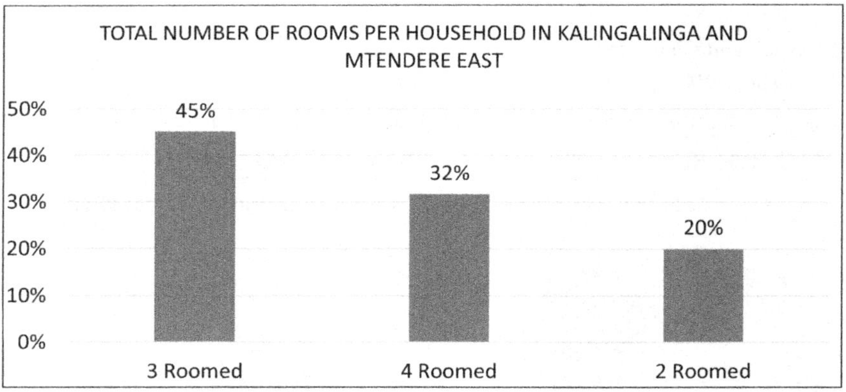

Figure 26. Showing Household size in Kalingalinga and Mtendere East

Source: Field, 2022

3.5.1.3 Level of Education for Household heads in Kalingalinga and Mtendere East

Most (37%) of the household heads had grade 9 education, whereas the second majority (33%) in the two settlements had grade 12 education. The rest (13%) are grade 7s, in tertiary, and (3%) grade 8s and (2%) grade 10s, as shown in figure 27.

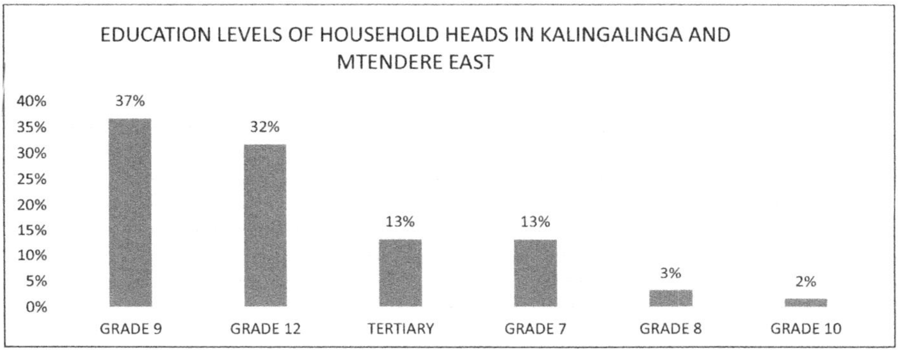

Figure 27. Showing level of education of household heads in Kalingalinga and Mtendere East

Source: Field Data, 2022

3.5.1.4 Occupation of Household Heads in Kalingalinga and Mtendere East

The most important (30%) occupation in the two settlements is trading followed by unskilled labour employment such as homecare (12%) security (10%), civil service (7%), vocational labour employment (5%), while the remaining fraction (2%) of the population fall in other occupations as illustrated in the figure 28.

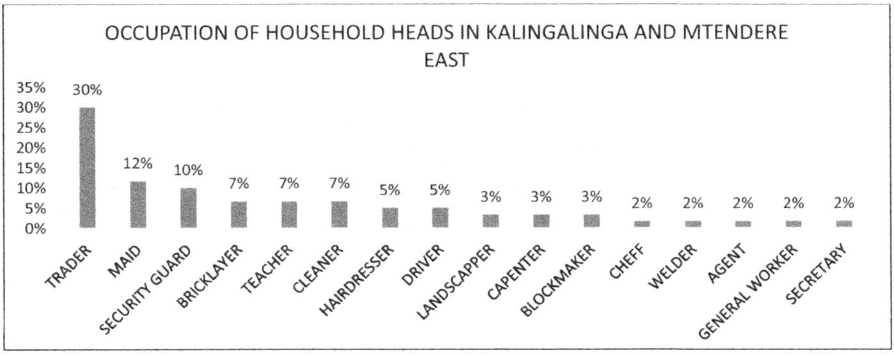

Figure 28. Showing Household size in Kalingalinga and Mtendere East

Source: Field Data, 2022

3.5.2 Existing Grassroot Initiatives for Resilient Housing in Kalingalinga and Mtendere East

Grassroot initiatives that are existent so far to foster resilient housing in the two settlements are that, all (100%) of the household heads had intentions of extending their houses, followed by their efforts (95%) to improvise, then their intentions to resell property (83%), then joining social cash transfer networks (72%) and involvement in community action (65%), least among them being the actual extending houses (43%) as shown in figure 29.

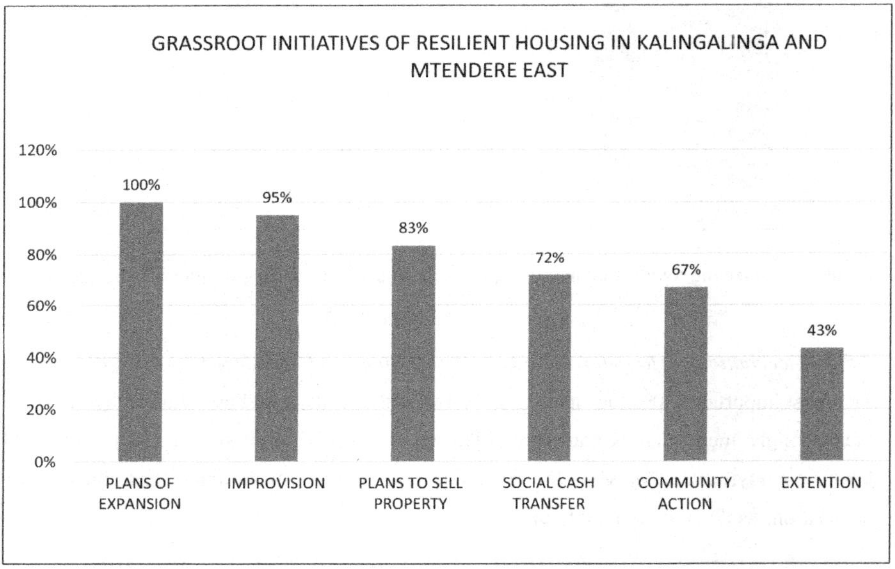

Figure 29. Showing Grassroot initiatives for resilient housing in Kalingalinga and Mtendere East

Source: Field Data, 2022

3.5.3 Potential for Enhanced Resilience of Housing in Kalingalinga and Mtendere East

According to a one-way ANOVA 2-sample variant F-Test at alpha level 0.5, one-tailed, findings reveal that, there is a high potential of enhancing resilient housing in Kalingalinga and Mtendere East. The test, results indicate an F-Value (1) greater than the F-Critical Value (0.198007), which entails that we reject the Null Hypothesis which states that, 'there is low potential for enhancing resilience in Kalingalinga and Mtendere East', while, we accept the

Alternative Hypothesis which states that, 'there is a high potential for enhanced resilience in the two settlements.' This is shown in table 7.

Table 7. Showing a One-Way ANOVA F-Test Two-Sample Variance Results for the potential of resilient housing in Kalingalinga and Mtendere East

HOUSING RESILIENCE POTENTIAL		HIGH	LOW
possession of valid occupancy license/land record		60	0
provision for upward adjustment		60	0
proximity to service lines (within 50 meters radius)		60	0
intentions of extension but limited resources		47	13
evidence of improvision		60	0
occupation in trade, skills and civil service		28	32
	Mean	52.5	7.5
	Variance	171.1	171.1
	Observations	6	6
	df	5	5
	F	**1**	
	P(F<=f) one-tail	**0.5**	
	F Critical one-tail	**0.198007**	

Source: Author, 2022

3.5.4 Grassroot Initiatives' Extent of Adaptability to Housing Challenges in the two areas

According to a one-way ANOVA 2-sample variant F-Test at alpha level 0.5, one-tailed, findings reveal that, the existing grassroot initiatives are not well adapted (not resilient enough) to meet the minimum housing standards of spatial planning in Kalingalinga and Mtendere East. The test results indicate an F-Value (1.911) less than the F-Critical Value (5.505), which entails that we accept the Null Hypothesis which states that, 'the existing grassroot initiatives are not well adapted to foster resilient housing in Kalingalinga and Mtendere East', while, we reject the Alternative Hypothesis which states that, 'the existing grassroot initiatives are well adapted to enhance resilient housing in the two settlements.' This is based on the multiple variables as shown in the table 8.

Table 8. Showing a One-way ANNOVA Two sample variance Test Results for the Standard of Housing Inspite of Grassroot Initiatives of resilience in Kalingalinga and Mtendere East.

HOUSING STANDARDS	NOT MET STANDARD	MET STANDARD
standard wall material over compromised	31	29
complete roofing over semi roofing	29	15
flushable toilet versus pit latrine	23	11
indoor tap water versus outdoor sources	53	7
1 person per room versus more than 1 person	59	1
prospective to sell over no prospective	37	23
Mean	38.667	14.333
Variance	203.867	106.667
Observations	6	6
df	5	5
F	**1.911**	
P(F<=f) one-tail	**0.247**	
F Critical one-tail	**5.050**	

Source: Author, 2022

3.5.5 The role of Spatial Planning in boosting Grassroot Initiatives for Enhanced Resilient Housing in the two settlements

The role of planning to boost grassroot initiatives is mapped out in a framework that interlinks initiatives and structures at grassroot level to formal planning structures. The mapping strategy has used an integrated planning approach that adopts the PPS to adapt spatial planning to grassroot initiatives as shown in figure 30.

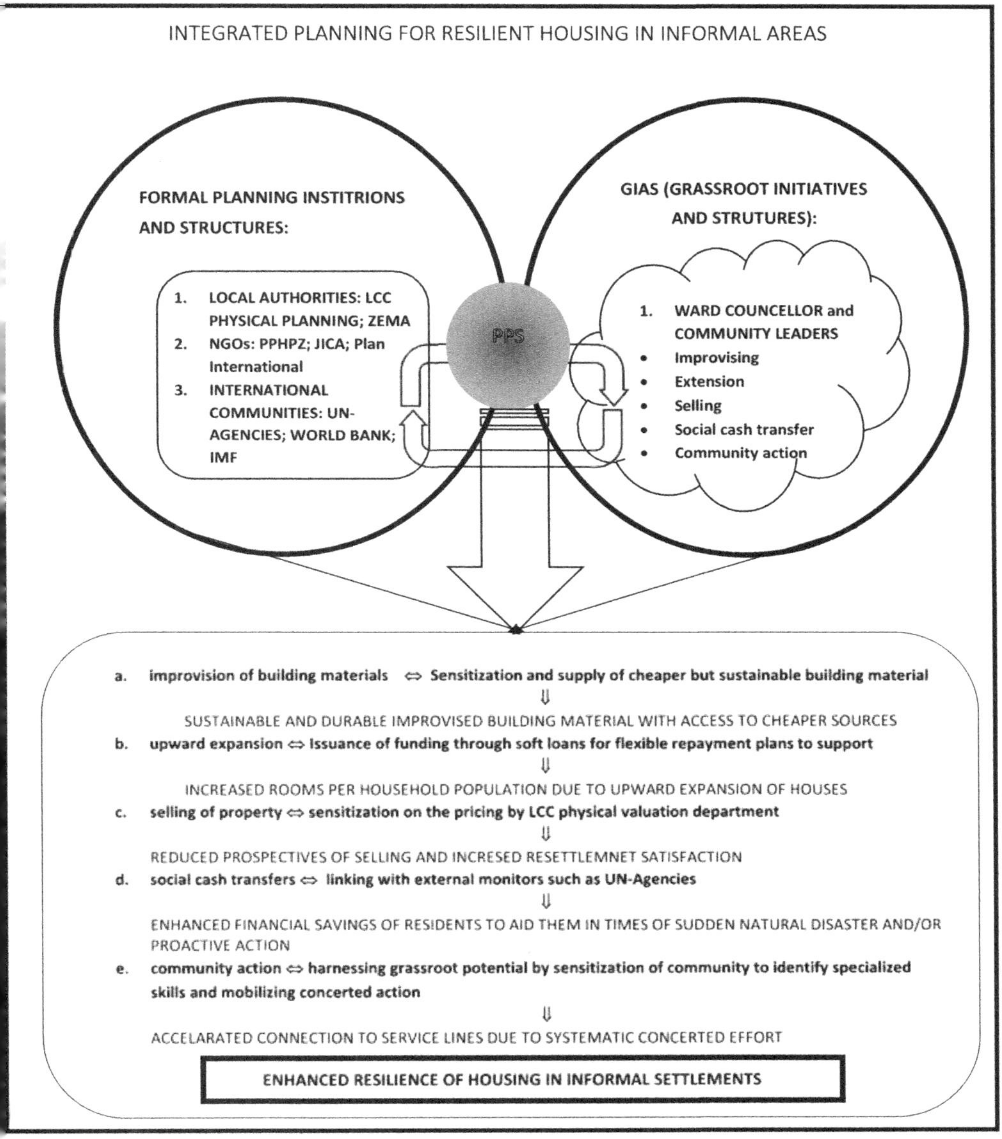

Figure 30. Showing Integrated Polycentric Planning model to harness Grassroot Initiative for enhanced resilient housing in informal settlements

Source: Adapted and Generated by Author, 2022

3.6 DISCUSSION

This section discusses the findings of the study by elaborating more on how existing housing crisis, it's potential for resilience and how planning clan plan a role to enhance resilience.

3.6.1 Respondent Characteristics

The ratio between male to female headed households is rather thin in Kalingalinga and Mtendere east, meaning both genders are proactive in running households. This entails that, women who were mostly occupied by SMEs ought to be supported to do more in sustaining their households. The average of 3 persons per room in a single house entailed overcrowding according to physical planning standards. This predicament calls for dire need to expand housing infrastructure. Meanwhile most dwellers of the two settlements do not sustain higher levels of education but at least have some elementary skills. Therefore, in order to boost their efforts to achieve resilient housing, planners have to condescend to their level of operation in order to integrate their initiatives with formal standards. Since trade is the most important economic activity occupying the dwellers, it becomes expedient that empowerment structures through project funding can be done to boost businesses. Using their income levels, soft loans with flexible payment plans can be arranged for the residents to help them increase their earnings.

3.6.2 Existing Grassroot Initiatives for Resilient Housing in Kalingalinga and Mtendere East

The grassroot initiatives aimed at fostering resilient housing in Kalingalinga and Mtendere East show how zealous the dwellers are to make their housing adaptable to prevailing physical, social and economic conditions in their settlements. All of them intend to extend their houses and would not mind whether they are extended upward or sideways. Besides that, almost all the houses were found with traces of mech-shift improvision such as cardboard on windows, stones on roofing sheets, cloth or plastic/sacs for doors, and mud as plastering material. Other initiatives were the willingness to sell properties for leverage out of low housing standards, involvement in social cash transfer networks, and deliberate but slow processes of extending their houses. Such efforts indicate that the people are not just watching their desperate housing crisis but have initiated means of adapting to the challenges. Therefore, it is important to recognize such efforts so that implementation of plans can have a more plausible basis. If planners and other helping agencies capitalize on the existing initiatives, strategies are likely to succeed because the dwellers own their initiatives and are likely to feel assisted rather than superimposed and dominated by outside forces.

According to the one-way ANOVA 2-sample variant F-Test at alpha level 0.5, 1 tailed, there is high potential of enhancing resilient housing in Kalingalinga and Mtendere East. Therefore, planners can take advantage of the residents' possession of valid occupancy licenses/land records, provision for upward adjustment, evident mech-shift improvisions, and proximity of the houses to service pipes. Using such existing potentialities, planners can have fertile ground to foster relevant action targeted at ensuring that these existing endowments are maximized. Also, planners can also take advantage of the dweller's occupations to link them to funding agencies to increase income and then marry this to their determination to extend their houses. This is likely to arouse interest and yield completion of most of their abandoned projects of extending their houses. Therefore, the two settlements present an opportunity where integrated planning can bring together informality and formality into a proliferated union to accomplish one good for habitable settlements.

3.6.3 Extent of Grassroot Initiatives' Adaptability to Housing Challenges in the two Settlements

A one-way ANOVA 2-sample variant F-Test at alpha level 0.5, one tailed, has also revealed that, the grassroot initiatives have not adequately adapted housing in the two settlements to meet minimum planning standards. Therefore, there is no significant improvement in the housing situation in spite of the initiatives and a high potential for resilience. This is seen in the compromised wall materials, high number of semi roofed houses, use of pit latrines, out-door sources of water, greater average number of persons per room, and higher prospective to sell despite not understanding the dynamics of informal land markets. Therefore, this raises a need for spatial planners to apply strategies that will foster resilience. Spatial planning can integrate formal planning systems, that have greater influence and social/legal and financial energy to grassroot initiatives, thereby boosting them up for improved implementation to enhance resilient housing in the two informal settlements.

3.6.4 The Role of Spatial Planning in boosting Grassroot initiatives for Enhanced Resilient Housing in the two settlements

The mapping strategy for integrated planning to boost grassroot initiatives has used an integrated planning approach that adopts the PPS. The logic is to devolve planning from higher levels to lower levels and make it more relevant and practical. This process is a dual flanked attack on the needs at both institutional and household level. The strategy the concomitant challenge that comes with formal institutions' negligence of informality. Through polycentricity, planning can be brought to the grassroot so that simple but practical initiates such as the intention to extend the house can be funded through a soft loan with flexible and affordable payment plans; social and economic planners linking those who intend to sell property with relevant agencies that help in valuation of property so that the sellers retain sufficient proceeds for leverage; also, local authorities through stakeholder meetings can combine grassroot initiatives with goodwill from NGOs such as PPPHPZ, JICA and the American and Swedish Government to both guide and boost local initiatives. Following the high potential for enhanced resilience in the two settlements, the PPS finds fertile ground to flourish and have its greatest positive impact on the lives of people in Kalingalinga and Mtendere East.

3.7 CONCLUSIONS

The study has shown that there are a number of initiatives at grassroot level striving to make housing as resilient as possible in Kalingalinga and Mtendere East. Among these include the intentions to sell properties, the improvision of inferior materials, the involvement in social cash transfers as saving mechanisms, the involvement in community action, and the actual extension of the houses. Inspite of such initiatives, the housing situation in the two areas has not adapted itself to meet minimum housing standards. Meanwhile, it has also been unveiled through statistical testing, that there is high potential for enhanced resilient housing in the two areas owing to the availability of contingencies such as the residents' intention to extend their houses, their possession of occupancy documents and sources of income, space for upward adjustments, as well as proximity to service pipes. Nevertheless, in spite of such high potential for enhancing resilience, a one-way ANOVA 2 sample variant F-Test at alpha level 0.5 showed that there was no significant improvement in the resilience of housing in the two settlements. This justified the need to innovate a planning approach that can harmonize key players so that the engines of implementation can run effectively at both institutional and grassroot level. Using such an opportunity, the study has further mapped out the connection between formal structures of planning with community structures and grassroot initiatives. This way, an integrated planning approach that synchronizes the grassroot efforts and institutional arrangements using the PPS was adopted and adapted to solve the missing links between the two. It is hoped that through such a unified approach, solutions with greater momentum for practicability and relevance can be reached to enhance resilient housing in informal settlements of Lusaka and other cities in the world facing similar challenges.

3.8 RECOMMENDATIONS

i. Through dialogue the LCC and Ministry of Finance and National Development Planning establish access to loans for extension of houses with flexible payment plans.

ii. PPPHPZ hold workshops and meetings with community leaders and conduct survey to find out the specific materials that the residents have improvised and link the people to cheaper sources of such materials. Also, through such interactions, can sensitize the residents, via media platforms on community initiatives/actions.

iii. JICA and the World Bank can partner with funding agencies to harness processes in the social cash transfer initiative to externally monitor the distribution and management of funds among the residents. They can also further make follow up surveys to evaluate the progress of entrepreneurship ventures of the beneficiaries.

iv. The LCC can engage the physical valuation department to evaluate properties and disseminate the information to prospective sellers of property. In the same vein, they can also advertise to prospective buyers and regulate the transactions.

ACKNOWLEDGEMENTS

I thank the Almighty God for the opportunity to write this paper and to make my submission to the AURH. Firstly, thanks to my friend and co-author Mr. Kaulule C., and my other co-authors Dr. Chibamba D., Dr. Ingram A., Dr. Siame G., and Dr., Nyanga P. I also thank my friend Miss. Chilufya K., for my perusing work. Specially thanking my young sister Fridah for perusing my work and assisting me with the laptop. I am grateful to the LCC, PPPHPZ and community leaders as well as the residents of Kalingalinga and Mtendere East for affording me the opportunity to conduct the surveys.

REFERENCES

[1] Akinola, R. S. (2016) *Urbanization, Collective Action and Coping Strategies in Informal Areas of African Cities: A Polycentric Environmental Planning Perspective.* In: Dynamics and Resilience of Informal Areas. pp5-24.

[2] Chibuye, M. (2014). Interrogating Urban Poverty lines: The case of Zambia. *Environment and Urbanization, International Institute for Research and Development.*, 26: 1, 236-256.

[3] Chikuta, E. 2016. *Implications of Informal Settlement Upgrading on Urban Development in the City of Lusaka: The Case of Kalingalinga.* A paper presented at the Urban Research Forum Maputo, Mozambique on 7th June 2016, pp4-7.

[4] Chirwa, M. (2019), *Informal land market dynamics in Kalingalinga settlement of Lusaka.* Master's dissertation. UNZA Repository, Lusaka.

[5] Chitengi, H. S. (2016). *Growth and resilience of informal settlements: toward informal contractual frameworks as alternatives for enhancing low-income housing, Housing and Society.* 43: 2, 138-155.

[6] Dobson, S., Nyamweru, H. and Dodman, D. (2015). Local and participatory approaches to building resilience in informal settlements in Uganda. *Environment & Urbanization International Institute for Environment and Development.* 27(2): 605–620.

[7] Government of the Republic of Zambia., 2011a. *Living Conditions Monitoring Survey Report 2006 and 2010*, Living Conditions Monitoring Branch, Central Statistics Office (CSO), GRZ, Lusaka, 304 pages.

[8] JCTR. (2010). *Satellite Homes Report. First Quarter*, Jesuit Centre for Theological Reflection (JCTR), Lusaka, 3 pages.

[9] Johannes, J. (2020). Land Rent Theory. In: Audrey Kobayashi (Ed): *International Encyclopaedia of Human Geography* (2nd Ed). Elsevier, Amsterdam. 8: 93-98.

[10] Jones, R. P. (2017), Housing Resilience and the Informal City. *Journal of Regional and City Planning.* 28: 129.

[11] Jones, R. P., and Suhartini, N. (2014). Reframing approaches to conceptualizing urban governance in Melanesia: Insights from Jayapura and Port Moresby. *Jurnal Perencanaan Wilayah dan Kota.* 25 (2), 96–114.

[12] Revell, K. (2010). *Working with Informality: Increasing Resilience in Cities of the Global South.* 46th ISOCARP Congress 2010, Kenya, Nairobi.

[13] Rivlin, L. G. (2007). *Found Spaces: Freedom of Choice in Public Life*. In: Franck K. A. and Q. Stevens (Eds), *Loose Space: Possibility and Diversity in Urban Life.* Routledge, London, pp38–53.

[14] Sudmeier-Rieux, K. (2014). Resilience -an Emerging Paradigm of Danger or of Hope? *Disaster Prevention and Management: An International Journal.* 23.

[15] Tinsley, E. (1997). *Mechanics of Informal Land and Markets, A theoretical Position.* Dissertation: Massachusetts Institute of Technology, pp50-61.

[16] UN Reports: *SDG 11 Indicators, 2021.*

Author's Biography

Mululu Chirwa is a qualified Spatial Planner and Educator of Geography with experience in field research. He is a holder of a MSc degree in Spatial Planning. He is a researcher and a writer with deep interest in civil engineering and spatial planning. He patronizes an Eco-Club where he inspires and cultivates innovation and technology to offer solutions to community challenges. Mululu is also passionate about reshaping informal landscape and improving the lives of the vulnerable in society. He has done research and written an article to help foster resilience in two informal settlements in the city of Lusaka. He is currently working to expand the Eco-Club into a more formidable organization to encompass stakeholders that can help merge ideas with practice to help informal communities. He is also the section head -Geography at Kabulonga Girls Secondary School of Lusaka. Mululu also won two awards of excellence in applied sciences at the University of Zambia during a postgraduate seminar presentation among a group of 84 participants. He also pioneered a team of students to win an award in a 'smart city' challenge competition organized by Plan International and Worldwide Fund

Publisher: Eliva Press Global Ltd

Email: info@elivapress.com

Made in United States
Orlando, FL
04 November 2026

90201146R00049